VOLUME ONE HUNDRED AND THIRTY ONE

ADVANCES IN COMPUTERS

VOLUME ONE HUNDRED AND THIRTY ONE

ADVANCES IN COMPUTERS

Edited by

ALI R. HURSON

Missouri University of Science and Technology, Rolla, MO, United States

Academic Press is an imprint of Elsevier
50 Hampshire Street, 5th Floor, Cambridge, MA 02139, United States
525 B Street, Suite 1650, San Diego, CA 92101, United States
The Boulevard, Langford Lane, Kidlington, Oxford OX5 1GB, United Kingdom
125 London Wall, London, EC2Y 5AS, United Kingdom

First edition 2023

Notices

Knowledge and best practice in this field are constantly changing. As new research and experience broaden our understanding, changes in research methods, professional practices, or medical treatment may become necessary.

Practitioners and researchers must always rely on their own experience and knowledge in evaluating and using any information, methods, compounds, or experiments described herein. In using such information or methods they should be mindful of their own safety and the safety of others, including parties for whom they have a professional responsibility.

To the fullest extent of the law, neither the Publisher nor the authors, contributors, or editors, assume any liability for any injury and/or damage to persons or property as a matter of products liability, negligence or otherwise, or from any use or operation of any methods, products, instructions, or ideas contained in the material herein.

ISBN: 978-0-443-21566-7
ISSN: 0065-2458

For information on all Academic Press publications
visit our website at https://www.elsevier.com/books-and-journals

Publisher: Zoe Kruze
Developmental Editor: Palash Sharma
Production Project Manager: James Selvam
Cover Designer: Greg Harris

Typeset by STRAIVE, India

Contents

Contributors

Asma Ben Letaifa
Mediatron Lab., SUPCOM, University of Carthage, Tunis, Tunisia

Phillip Bradford
Department of Computer Science and Engineering, University of Connecticut, Storrs, CT, United States

Harry J. Foxwell
George Mason University, Fairfax, VA, United States

Poulami Majumder
Maulana Abul Kalam Azad University of Technology, West Bengal, India

Hela Makina
Mediatron Lab., SUPCOM, University of Carthage, Tunis, Tunisia

Prasanna Mani
SITE, VIT, Vellore, India

Arul Treesa Mathew
SITE, VIT, Vellore, India

Abderazek Rachedi
Computer Science Lab. (LIGM UMR8049), University Gustave Eiffel, CNRS, Marne-la-Vallée, France

Partha Pratim Ray
Department of Computer Applications, Sikkim University, Gangtok, India

Marcus Tanque
Independent Researcher, Washington, DC, United States

Q.M. Jonathan Wu
School of Computer Science, Engineering Research Center of Digital Forensics by Ministry of Education, Nanjing University of Information Science and Technology, Nanjing, China; Department of Electrical and Computer Engineering, University of Windsor, Windsor, ON, Canada

Sheng Wu
Operating Unit on Policy-Driven Electronic Governance, United Nations University, Guimarães, Portugal

Chengsheng Yuan
School of Computer Science, Engineering Research Center of Digital Forensics by Ministry of Education, Nanjing University of Information Science and Technology, Nanjing; Key Laboratory of Public Security Information Application Based on Big-Data Architecture by Ministry of Public Security, Zhejiang Police College, Hangzhou, China

Qianyue Zhang
School of Computer Science, Nanjing University of Information Science and Technology, Nanjing, China

Preface

Traditionally, *Advances in Computers*, the oldest series to chronicle of the rapid evolution of computing, annually publishes four volumes, each one typically comprised of four to eight chapters, describing new developments in the theory and applications of computing.

Volume 131 is an eclectic volume inspired by recent issues of interest in research and development in computer science and computer engineering. The volume is a collection of six chapters as follows:

Due to the advances in information, communication, and computing technologies, the healthcare industry has undergone massive changes. New services and new expectations are becoming parts of common and traditional practices. How to maintain the quality of services in the face of massive information generated on a daily basis while guaranteeing the privacy of individuals and maintaining the security of personal medical information are becoming new issues that need to be addressed in this newly evolved industry. In response to the rapid changes in the healthcare industry, Makina et al. in Chapter 1 entitled "eHealth: Enabling technologies, opportunities and challenges" survey and articulate the relevant use cases in the healthcare industry. The chapter identifies the technical and Quality of Service requirements of each use case and surveys recent work in the enabling technologies. Finally, the challenges and open research issues as reported in the literature are discussed.

Cancer research is becoming more and more diversified. Different techniques and approaches are being developed and adopted. Application of blockchain technology to provide a pathway, interconnectivity, and interoperability among different cancer research centers is the main theme of Chapter 2. In this chapter entitled "A review on cancer data management using blockchain: Progress and challenges," Ray and Majumder present a holistic approach to care, cure, and manage cancer by application of blockchain technology. The chapter argues about the usefulness of blockchain for cancer care and research and delves into the current trend in blockchain-based cancer care. Finally, a blockchain model called BCCCANA for cancer diagnosis, treatment, cure, and management is introduced.

Cybersecurity solutions, i.e., policies, procedures, techniques, and frameworks for mitigating long-term threats, risks, and potential vulnerabilities, are becoming of major interest and concern in various organizations.

In Chapter 3 entitled "Cyber risks on IoT platforms and zero trust solutions," Tanque and Foxwell discuss Zero Trust and Internet of Things device security, protection, and interoperability. They outline how Zero Trust concepts and methods focus on the security issues related to IoT systems. The chapter further discusses details about current Zero Trust solutions for IoT security. Finally, future solutions, policies, and frameworks are addressed.

Chapter 4, entitled "A realtime fingerprint liveness detection method for fingerprint authentication systems" by Yuan et al., as the name suggests, focuses on fingerprint authentication and fingerprint liveness detection as a prerequisite and essential step before fingerprint authentication. The chapter proposes a three-step lightweight and real-time fingerprint liveness detection technique. Experimental results with five public fingerprint datasets are presented and compared against some other advanced models, based on metrics such as speed, size, and performance.

Fog and Edge computing platforms are the major theme of discussion in Chapter 5. In this chapter entitled "Collaborating fog/edge computing with industry 4.0—Architecture, challenges and benefits," Mathew and Mani introduce the concept of edge and fog computing and discuss the architecture and major components of each platform. The benefits and research challenges of each platform is articulated. Finally, the chapter has a comprehensive discussion about the application of edge and fog computing platform in the 4th Industrial Revolution.

Finally, in Chapter 6 entitled "Virtual Raspberry Pi-s with blockchain and cybersecurity applications," Tanque and Bradford question running cybersecurity applications on small and constrained virtual Raspberry Pi-s. The chapter explores cybersecurity applications on virtual Raspberry Pi-s augmented with blockchains and investigates whether characteristics of virtual machines such as redundancy, replication, cloning, and agility can benefit cybersecurity applications on small devices augmented with blockchain technology.

I hope that readers find this volume of interest and useful for teaching, research, and other professional activities. I welcome feedback on the volume as well as suggestions for topics for future volumes.

Ali R. Hurson
Missouri University of Science and Technology,
Rolla, MO, United States

CHAPTER ONE

eHealth: Enabling technologies, opportunities and challenges

Hela Makina[a], Asma Ben Letaifa[a], and Abderazek Rachedi[b]
[a]Mediatron Lab., SUPCOM, University of Carthage, Tunis, Tunisia
[b]Computer Science Lab. (LIGM UMR8049), University Gustave Eiffel, CNRS, Marne-la-Vallée, France

Contents

Advances in Computers, Volume 131
ISSN 0065-2458
https://doi.org/10.1016/bs.adcom.2023.04.001

Abstract

Recently, the healthcare industry has undergone rapid transformation from the traditional hospital-based scheme to a patient-centered approach. Patients are demanding newer, more personalized, and accessible healthcare services matching the speed of modern life. Remote monitoring, teleconsultation, telesurgery, robotics for assisted living, to mention a few, are examples of new medical services that are becoming widespread thanks to the advancements in Information and Communication Technologies (ICT) such as IoT wearables, cloud computing, reliable wireless networks, etc. However, these services produce massive amounts of data and need very strict QoS requirements related to latency, energy efficiency, data rates, mobility, positioning accuracy, security, privacy, etc. For the latency and energy efficiency requirements, Edge/Fog computing combined with AI/ML approaches are the answer to compensate the limitations of the cloud. For high data rates, mobility management, and positioning accuracy, the 5G mobile standard offer promising features to tackle the different use cases' needs. Security and privacy measures can be implemented in all the above-mentioned technologies. In this article, we review and describe the most relevant use cases in the healthcare industry. We identify the technical and QoS requirements of each use case. Then we survey recent works about the enabling technologies. Finally, we identify the challenges and open issues in the existing literature.

Abbreviations

5G-PPP	5G public-private partnership
6LoWPAN	IPv6 low power wireless personal area networks
BLE	bluetooth low energy
CPS	cyber physical system
EHR	electronic health records
eMBB	enhanced mobile broadband
FN	fog node
GHO	Global Health Observatory
IoT	internet of things
KPI	key performance indicator
LoRaWAN	long-range WAN
MCPS	medical cyber physical system
mMTC	massive mobile type communication
NFV	network function virtualization
NIH	National Institutes of Health
NIST	National Institute of Standards and Technology
NS	network service
PCC	patient-centered care
RPM	remote patient monitoring
SDN	software-defined networking
SLA	service level agreement
UAV	unmanned aerial vehicles
uRLLC	ultra-reliable low latency communication
WBAN	wireless body area network
WHO	World Health Organization

1. Introduction

Nowadays, we are living in the era of Healthcare 4.0 inspired from the term Industry 4.0. The latter is a concept that reflects the shift to automation, digitalization, personalization, and virtualization introduced in all industrial domains. Similarly, Healthcare 4.0 followed the same trends and has passed through four generations from Healthcare 1.0 to Healthcare 4.0 [1]. Early healthcare services were based on a hospital-centric scheme where everything was treated manually. Then, it was the decades between 1970 and 1990 that witnessed the beginning of digitalization by the emergence of Information and Technology (IT) systems and Healthcare 1.0 was introduced. So, doctors moved from handwritten notes to recording patient information into computers to be processed and stored. Next came Healthcare 2.0 in the following decade pushed by IT networks and the introduction of Electronic Health Records (EHR). Then by 2005, Healthcare 3.0 emerged with the integration of wearables and implants in health control. Additionally, EHR systems became networked and more developed to give access to every interferer when needed. Today, we are seeing the emergence of Healthcare 4.0 driven by precision medicine, IoT devices, advanced robotics, cloud computing, big data, and artificial intelligence.

With such new technologies, the hospital-based traditional model of healthcare turns to a distributed Patient-Centered Care (PCC) model. The Institute of Medicine defined PCC in [2] as: "Healthcare that establishes a partnership among practitioners, patients, and their families (when appropriate) to ensure that decisions respect patients want, needs, and preferences and that patients have the education and support they need to make decisions and participate in their own care". This does not mean that there is no need for hospitals or clinics, but healthcare is provided through a decentralized model where all interferers coordinate through ICT.

In Healthcare 4.0, several terms such as eHealth (electronic Health), mHealth (mobile Health), telemedicine and teleHealth can be used interchangeably although there is a tight difference between them. In the following, we will give the exact definition of each concept:

- eHealth was defined by the World Health Organization (WHO) as [3]: "the cost-effective and secure use of information and communications technologies in support of health and health-related fields, including health-care services, health surveillance, health literature, and health education, knowledge and research".

- mHealth, according to the same organization (WHO), is "the use of mobile and wireless technologies to support the achievement of health objectives" [4]. Similarly, the National Institutes of Health (NIH) defines mHealth as "the use of mobile and wireless devices (cell phones, tablets, etc.) to improve health outcomes, health care services, and health research" [5].
- Telemedecine is "the delivery of health care services, where distance is a critical factor, by all health care professionals using information and communication technologies for the exchange of valid information for diagnosis, treatment and prevention of disease and injuries, research and evaluation, and for the continuing education of health care providers, all in the interests of advancing the health of individuals and their communities". [6]
- TeleHealth, according to the Global Health Observatory (GHO) is: "the delivery of health care services, where patients and providers are separated by distance. Telehealth uses ICT for the exchange of information for the diagnosis and treatment of diseases and injuries, research, and evaluation, and for the continuing education of health professionals" [7] (Fig. 1).

A patient-centric eHealth architecture needs a multi-layer environment: intelligent devices to collect medical data, network infrastructure for data transmission and efficient datacenters for data processing and storage.

In the remainder of this document, the term eHealth will refer to all ICT-based healthcare services. eHealth services include remote health monitoring systems, providing medical care at home, enabling remote surgeries, encouraging proactive healthcare, fitness, and wellness programs, elderly care, treatment and medication at home, reducing wastage of pharmaceuticals due to accidental expiry, ensuring service continuity and remote monitoring even in the case of disaster areas (wars, earthquakes, volcanos, floods, etc.) or areas with poor coverage (desert, mountain, etc.) [8].

Managing a patient data is based on EHRs which are an essential part of eHealth infrastructure. According to Ref. [3], EHRs are defined as "real-time, patient-centered records that provide immediate and secure information to authorized users. EHRs typically contain a patient's medical history, diagnoses and treatment, medications, allergies, immunizations, as well as radiology images and laboratory results".

EHRs offer many benefits. They propose a centralized scheme including patient's medical history, diagnoses, medications, treatment plans, allergies, radiology images, and laboratory and test results (Fig. 2). Besides, they are generated and maintained by authorized providers in a digital and

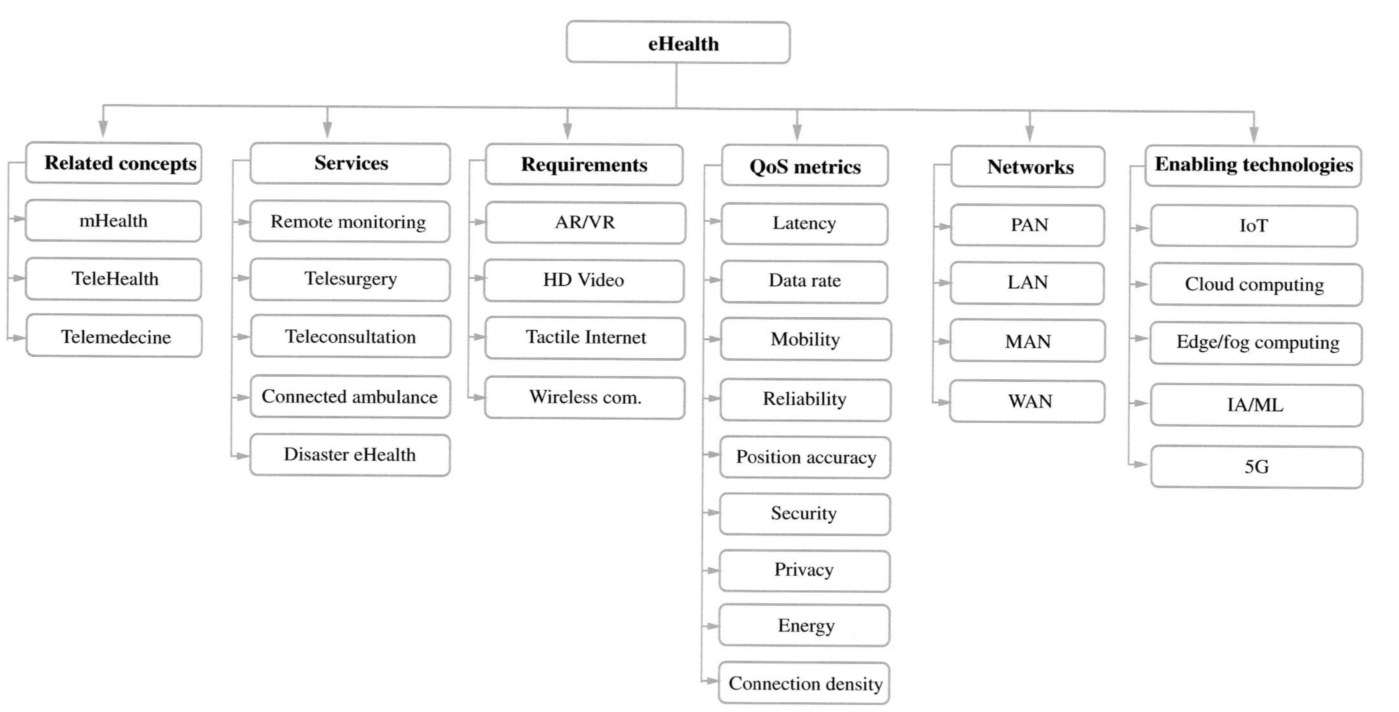

Fig. 1 Taxonomy of eHealth, its requirements and technologies.

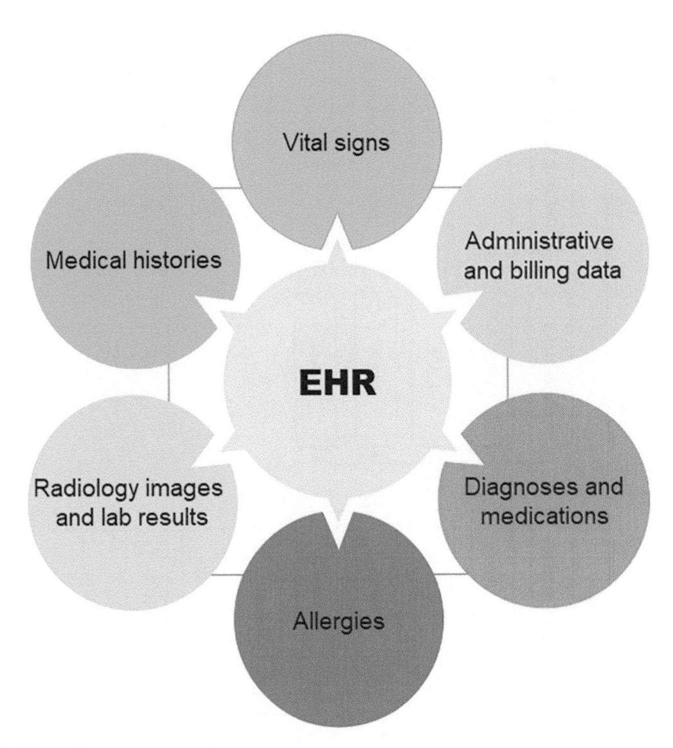

Fig. 2 EHR's main components.

standardized format that can be shared with other interferers which reduces transcription, access, and storage costs. Hence, they improve patient quality of care through better ailment management and patient education. EHRs can be accessed by different stakeholders [9]: medical staff (physicians, nurses, laboratory technicians, radiologists, etc.), technical experts (data scientists, computer engineers, implementation experts, etc.), decision-makers (medical administrators, regulatory agencies, etc.), patients, their care takers and families, insurance companies, etc.

1.1 Comparison with related surveys

There have been many surveys on eHealth that studied the subject from various aspects such as IoT, cloud computing, Edge/Fog computing, Artificial Intelligence (AI) and Machine Learning (ML), 5G networks, Security, etc. In Table 1 we give a review of different research related to the topic. IoT and cloud computing are considered as the two fundamental cornerstones for eHealth use cases. In Ref. [10], the authors give an exhaustive overview

Table 1 Previous surveys on eHealth.

		Technologies						
Refs.	**Topics covered**	**IoT**	**CC**	**EC/FC**	**AI/ML**	**5G**	**LPWAN**	**Security**
[9]	A survey on ML-applied algorithms to preserve security and privacy in healthcare		✓	✓	✓			✓
[10]	A survey on IoT-based eHealth services	✓	✓					
[11]	Opportunities and security challenges of cloud-based eHealth	✓	✓					✓
[12]	A survey on security of IoT-cloud-based eHealth	✓	✓					✓
[13]	Security and privacy challenges in IoT-cloud-based e-health systems	✓	✓					✓
[14]	A survey on fog computing opportunities and challenges for healthcare 4.0	✓	✓	✓				
[15]	Promises and challenges of IoT and fog computing in eHealth	✓	✓	✓				
[16]	A Survey on edge intelligence and IoT in Healthcare	✓		✓	✓			
[17]	A survey on smart edge computing in healthcare systems		✓	✓	✓	✓		
[18]	A survey on edge-cloud computing and AI in Internet of Medical Things	✓	✓	✓	✓			
[19]	Trends in IoT based solutions for health care: moving AI to the Edge	✓		✓	✓			
[24]	A survey on privacy provision in eHealth with Attribute-Based Encryption		✓					✓

Continued

Table 1 Previous surveys on eHealth.—cont'd

		Technologies						
Refs.	**Topics covered**	**IoT**	**CC**	**EC/FC**	**AI/ML**	**5G**	**LPWAN**	**Security**
[25]	Opportunities of LPWAN and embedded ML as enablers for the next generation of wearable devices				✓		✓	
[20]	A review on 5G network technologies for smart IoT-healthcare deployment	✓				✓		
[21]	Design and implementation of eHealth use cases on 5G network	✓				✓		
[22]	A study on the performance requirements of advanced healthcare services over cellular networks					✓		
[23]	A survey on communication requirements of healthcare applications over 5G networks	✓				✓		
This work	A review on emerging services, QoS requirements, enabling technologies and open issues in eHealth	✓	✓	✓	✓	✓	✓	✓

CC: Cloud Computing, EC: Edge Computing, FC: Fog Computing

on IoT-based eHealth systems, platforms, and applications. In Ref. [11,12,13], the authors survey cloud-based eHealth architectures and their challenges regarding security and privacy. In Ref. [14,15], fog computing is studied as a key layer that enhances Quality of Service (QoS) in eHealth in terms of low latency, reliability, and security. AI/ML coupled with Edge/Fog computing are surveyed in Refs. [9,16,17,18,19] to ensure smart processing of medical data for diagnosis, prevention, QoS provision and security purposes. 5G mobile networks are also considered as a promising enabler for healthcare services as the offered capabilities can meet eHealth QoS requirements. They have been the subject of several surveys [20,21,22,23] that focus mainly on medical QoS requirements. Security is also considered as a major challenge that must be considered in the different layers of an eHealth system/application. Multiple propositions are given to ensure secure transmission, processing, and data storage [24].

Our contribution is to provide a review on the different enabling technologies for eHealth including:

- A taxonomy for emerging eHealth services and their technical and QoS requirements.
- A review of recent studies on enabling technologies such as IoT, cloud computing, Edge/Fog computing, AI/ML, and 5G networks to achieve the QoS requirements of the main eHealth services.
- Open research directions and issues.

1.2 Organization of the paper

The structure of the survey is as shown in Fig. 3. The remaining part of this article is organized as follows. In Section 2, we introduce the typical

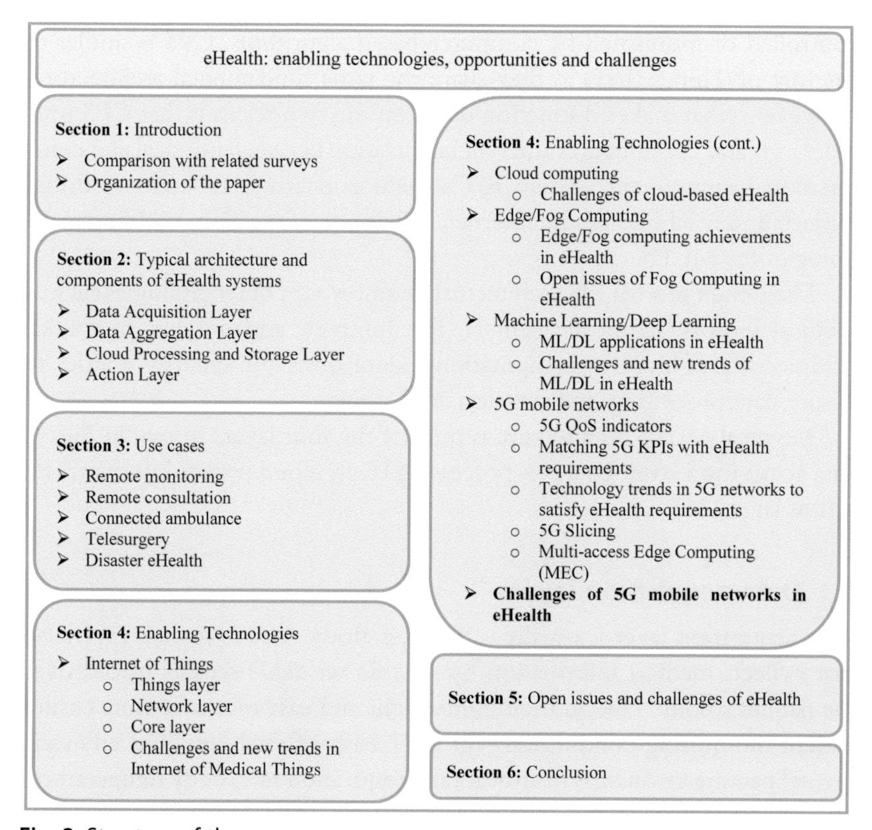

Fig. 3 Structure of the survey.

architecture of an eHealth system and its components. In Section 3, we describe a set of eHealth use cases to identify their QoS requirements. Section 4 details enabling technologies of eHealth and identifies their challenges. Following this, in Section 5, we summarize the open issues that must be studied to strengthen eHealth systems. Finally, Section 6 concludes the paper.

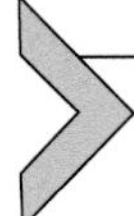

2. Typical architecture and components of eHealth systems

eHealth system so-called Medical Cyber Physical System (MCPS) incorporates computational and networking techniques with medical and physical implementations to satisfy a medical purpose (monitoring of vital signs, remote surgery, etc.). MCPS is derived from the general term Cyber Physical System (CPS) which refers to every physical mechanism controlled or monitored by computer-based algorithms. CPS is similar to Internet of Things (IoT) as they share the same fundamental architecture. However, what makes distinction between the two terms is that CPS provides a greater combination and synchronization between physical and computational components where IoT's main concern is to connect things (objects and machines) to the internet. IoT is one of the main enabling technologies for MCPS.

The design of a MCPS architecture requires a set of technologies such as medical sensors, cloud datacenters, fast Internet, and wireless networks. Additionally, it involves computational algorithms and security politics to ensure data processing, transmission, and storage.

A typical MCPS architecture is built of the four layers shown in Fig. 4: data acquisition layer, data pre-processing layer, cloud processing layer, and action layer [26].

2.1 Data acquisition layer

Data acquisition layer is usually a Wireless Body Area Network (WBAN) that collects medical information by specific wearable sensors placed over the patient's body. Due to their lightweight and ease of use, sensors ensure patient monitoring continuously on 24/7 basis of vital signs and environmental parameters such as heartbeat rate, respiration rate, body temperature, blood glucose, body movement, hydration level, etc. However, sensors have low computational and storage power, so they relay the collected data to

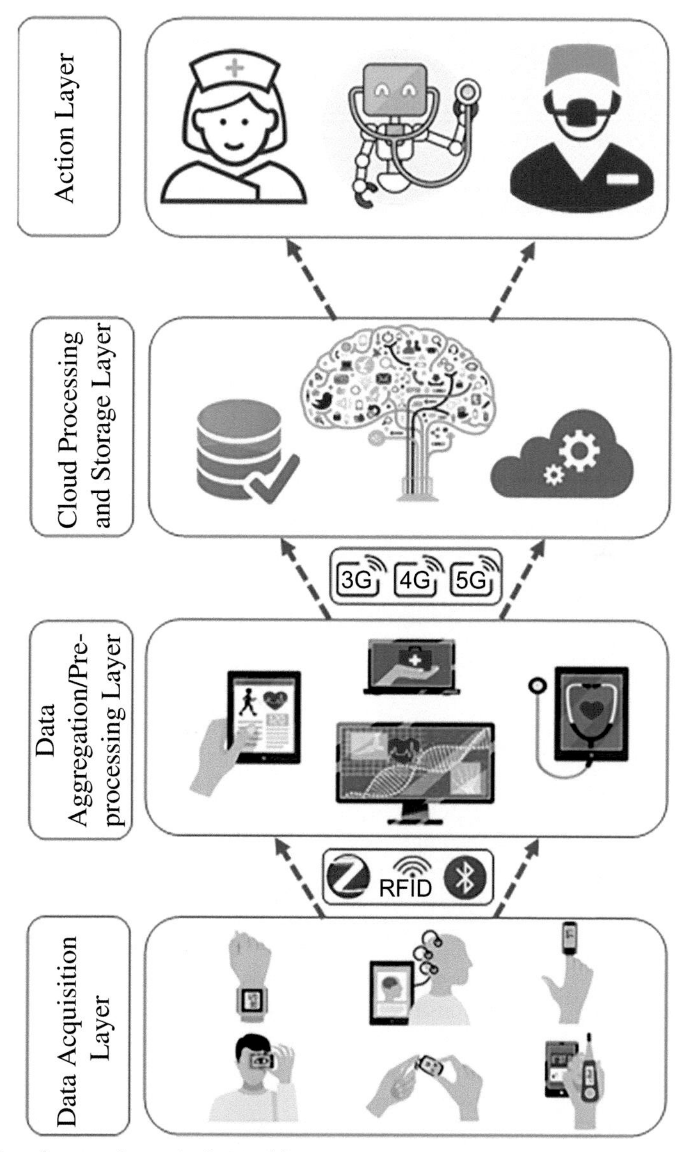

Fig. 4 Four layers of a typical eHealth system.

powerful devices to be processed. Different technologies can be used for data transmission such as Bluetooth, ZigBee, Wi-Fi, etc. In Table 2, we collected several systems for remote monitoring using different network technologies to relay the sensed data to be processed in upper layers.

Table 2 Communication technologies for different remote monitoring applications.

Refs.	Purpose	Used sensors/ devices	Measured parameters	Communication technology
[15]	Heart Rate monitoring	– Smart eyeglasses	– Heart rate	BLE
[15]	Remote monitoring for Parkinson's disease patients	– Smart glove	– Motor symptoms in the hand (tremors, rigidity/ slowness of movements)	BLE
[27]	Remote monitoring for special cases (pregnant women for example)	– iBeacon Bluetooth Low Energy BLE 4.0 Proximity Devices – Smartphone	– Blood pressure – Oxygen level	BLE/ 5G
[28]	Elderly monitoring	– Mysignals HW V2 platform	– Oxygen in blood – Blood glucose – Body temperature – Blood pressure – Patient position	Wi-Fi
[29]	In-home smart sensing	– Gas sensors – Accelerometer module – Sleeping monitor module	– Gases concentration in the breathed air – Individual's motion (fall detection)	Not specified
[30]	Elderly/chronic patient monitoring	– ECG – Blood pressure monitor – Pulse OxyMeter – Mobile phone	– Blood pressure – Body temperature – Electrocardiograms – Oxygen saturation	6LoWPAN
[31]	Elderly monitoring	– ECG – GPS	– Beat rate – Location (to detect falling events)	ZigBee
[32]	Chronic patient monitoring	– ECG – Accelerometer and pedometer (to detect the patient's activity) – Location sensor – Thermometer	– Beat rate – Patient's state and activity (standing, walking, running) – Location	Bluetooth/5G

Table 2 Communication technologies for different remote monitoring applications.—cont'd

Refs.	Purpose	Used sensors/devices	Measured parameters	Communication technology
[33]	A six-minute walk test (6MWT)	– A bracelet tag – Cardiac monitor	– Heart rate – Individual's activity	RFID/BLE
[34]	Real-time Heart Attack Mobile Detection	– Smart watches	– Voice control – Gesture control	WiMax/4G

2.2 Data aggregation/pre-processing layer

To cope with the limited capabilities of collecting sensors, an intermediate device between acquisition layer and the cloud is needed. Depending on the adopted architecture, this layer may consist in an IoT gateway, a fog node, or an edge device having the computational and storage power to concentrate data, perform processing tasks and sometimes short or mid-term storage. It can be a dedicated concentrator, a smartphone, or a computer with a dedicated internet connection.

2.3 Cloud processing and storage layer

The cloud layer performs two main functions: data processing and storage. It hosts medical applications and algorithms to analyze the collected data and give decisions and predictions about patient's situation. Furthermore, it provides long term storage in enhanced performance and efficient energy datacenters.

2.4 Action layer

The action undertaken in this layer may be "active" or "passive" [26]. In the first case, the process run in the cloud turns a response into an actuator such as a robotic arm. Robot-assisted surgery is a use case corresponding to such an action [35]. In the second case, there is no concrete action undertaken. Elderly monitoring is an example of a passive action [28].

The above-mentioned layers may be implemented with different technologies and protocols. In the following sections, we give an overview on the main supporting technologies.

3. Use cases

While going through the surveyed research works, a wide range of healthcare services were proposed. We grouped the most promising ones into five categories according to their social impact, shared architecture, and their technical/QoS requirements. These use cases are: remote monitoring (both in hospital and at home, specifically elderly monitoring), telesurgery, remote consultation, connected ambulance and disaster eHealth. It is worth noting that some of these use cases are related to each other's and may share key technologies. For example, telesurgery and connected ambulance may include some scenarios of remote monitoring. In the following sub-sections, we describe these scenarios in more details.

3.1 Remote monitoring

Remote patient monitoring is considered as a key factor for more efficient and proactive healthcare delivery especially for chronic diseases and elderly people. Patient vital signs and environmental parameters can be collected and analyzed using medical sensors and devices without any need to travel to health centers and have a face-to-face appointment with a medical professional. The typical requirements for such use case are specialized sensors, implants or wearables that collect vital signs, physiological and environmental parameters and send them to a smartphone application. The latter serves as a gateway and principal collector for many and heterogeneous data, see Fig. 5. In case of abnormal detected values, a notification is sent to a supervising agent [46,47,48]. Medical data are collected in a specified frequency.

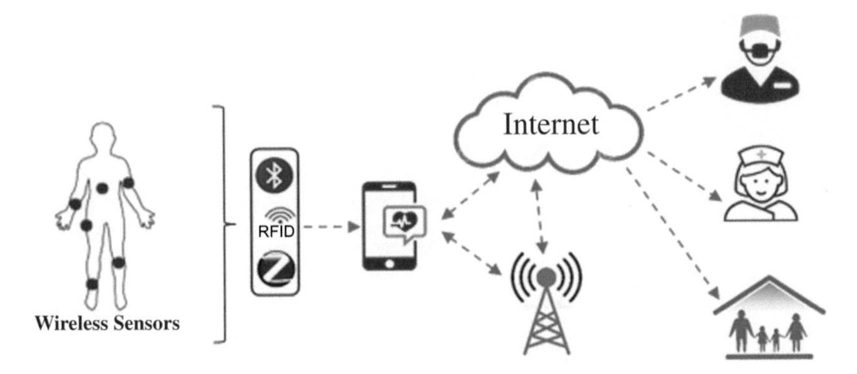

Fig. 5 Remote monitoring.

Then they must be processed to extract meaningful information about the patient's state and to take decisions and predictions based on ML and data mining algorithms. Remote monitoring can be performed either in hospital or at home. Particularly, elderly monitoring is seeing a great success and various applications.

3.1.1 In-hospital remote monitoring

In this case, continuous monitoring is provided simultaneously to many in-hospital patients. Medical data are collected from both routine and sophisticated devices then examined in real-time to maintain the current health condition of the patient. For example, patients in the intensive care unit can be monitored continuously by measuring their vital signs such as blood pressure, oxygen level, heartbeat rate, etc., to predict probable heart attacks [22]. According to [36], in-hospital pervasive monitoring produces traffic flow with data rates about 1 Mb/s/patient. Thus, for a high number of observed patients in the limited area of the hospital, the resulting traffic can reach 300 Mb/s, while expected latency average is about 250 ms [22].

3.1.2 At-home remote monitoring

With a large variety of wearables and implants, new trends for at-home remote monitoring services are emerging including chronic disease monitoring, fitness tracking and health development, aid for the physically impaired, etc. [23]. A new generation of cheap and reliable smart clothing (shoes, socks, gloves, etc.) and accessories (eyeglasses, watches, bracelets, etc.) can measure both vital and physiological signs with easy and user-friendly interfaces. Besides, a smart house can be equipped with environmental sensors to measure ambient temperature, oxygen level, humidity, light exposure data, etc. At-home remote monitoring scenarios require reliable communication networks to handle the expected high number of users moving in different indoor/outdoor environments at different speeds.

3.1.3 Elderly monitoring

As life expectancy have been extended over the last years, new age-related healthcare demands have arisen. Indeed, for elderly persons, the risk of sudden falls, stroke, brain injuries and neuro degenerative diseases increases. Besides, chronic patients living independently need remote assistance for medication management and emergency detection (falls, fires, etc.). Emerging elderly monitoring services are based on mobile applications gathering information from medical devices and environmental sensors to detect vital signs and physiological activities like body temperature, heartbeat rate, activity levels, bathroom use, eating habits, sleep patterns, etc. For any detected abnormality, a personalized report can be sent to a patient's relative, or an urgent alert is raised to emergency service.

To ensure continued and smart monitoring, medical sensors require to have a small size design to be carried easily, low power consumption, high efficiency and reliability and to be always "on" and connected [37].

QoS requirements vary according to the applications. Reference [23] gives detailed QoS requirements for each application. For example, maximum latency tolerable for chronic disease monitoring and emergency event detection is about 50 ms. It must not exceed 1 s for vital signs monitoring, while it can reach a few seconds in the case of fitness tracking devices. Requirements for Bit Error Rate (BER) depends also on the application and often ranges between 10^{-10} and 10^{-5}. Energy efficiency is a critical factor in the case of remote monitoring. The needed battery lifetime can range from a few days for wearable devices to a few years for implants. Moreover, a connection density of up to 10,000 devices/km^2 and a traffic density exceeding 50 Gbps/km^2 are needed to ensure remote pervasive monitoring through mobile networks.

3.2 Remote consultation

For patients in rural or underdeveloped areas where only general physicians are available, there is often a lack of access to specialist clinicians. This challenge can be overcome through remote consultation which reduces the burden and the cost on the patient and decreases the contamination risk such in COVID-19 pandemic. In this case, a high bandwidth network connectivity is required to enable video interaction between the patient (and the local doctor) and the remote specialist. Besides, wearable devices are needed to collect patient's vital signs. It is expected that the typical number of IoT devices needed per appointment ranges between 1 and 10 [38]. For real-time video streaming, experienced data rates can reach 1 Gb/and acceptable latencies must not exceed 150 ms [22]. Besides, remote consultation needs secure data transmission, and the capacity to store and archive analysis sessions for possible review in later date.

3.3 Connected ambulance

In emergency situations, minutes or even seconds can be vital. That is why the "connected ambulance" [39,40] is defined to improve patient outcomes and avoid severe complications. In this case, patient information (such as

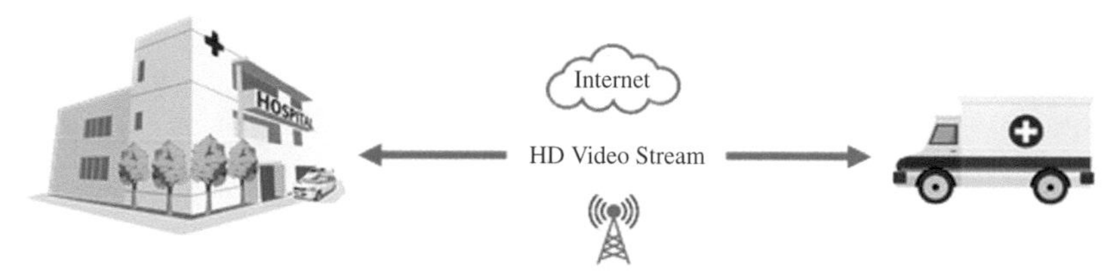

Fig. 6 Data exchanging in connected ambulance.

temperature, blood pressure, heart rate, etc) is collected through wearables, sensors, or streaming video cameras and transferred to hospital staff while the patient is being transported, see Fig. 6. Specialists can guide paramedics in the ambulance to perform emergency procedures before arriving to the hospital. In such scenario, the major challenges are related to high-definition video transmission and high reliability to maintain communication in the case of high mobility (up to 100 km/h). According to Ref. [23], maximum tolerable latency is around 10 ms for haptic feedback and around 250 ms for vital signs transmission.

3.4 Telesurgery

Telesurgery was first defined to refer to remote surgical procedures performed by robotics using computer technology and wireless networking [35]. This could also be extended to include collaboration between distant doctors to perform surgeries: a distant specialist can use an Augmented Reality/Virtual Reality (AR/VR) headset to watch a surgery in real time executed by another colleague to comment or even help him based on his own experience as shown in Fig. 7.

In the case of robot-assisted telesurgery, a distant surgeon equipped with a tactile console monitors and guides a mechanical arm of a robot to perform the surgery through ultra-reliable and low latency communication scheme [41]: after determining the location of the incision, the mechanical arm, which is fitted with a HD camera and other surgical instruments, cuts, stops bleeding, and sutures the patient. The surgeon-robot interaction involves tactile devices for high precision positioning control, HD camera for real-time video transmission and medical sensors for remote monitoring of the patient's state.

Because telesurgery involves different communication streams (camera flows, vital signs, real time haptic feedback, etc.), it has strict QoS requirements regarding latency, data rate, packet loss, availability, and reliability. In Refs. [23], the authors surveyed various experimental and simulation-based studies addressing quantitative QoS requirements for remote robotic-assisted

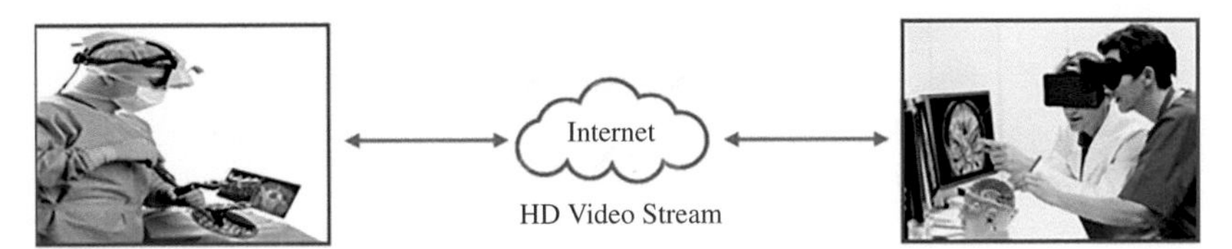

Fig. 7 Collaboration between distant doctors in telesurgery.

surgery. In particular, latency was identified as the most relevant indicator, and the surgical performance deteriorates in an exponential way as the latency increases. Depending on the difficulty of performed tasks during telesurgery, the used equipments, and the simulation environments, the identified upper bound of latency varies from 1 ms for haptic feedback to 700 ms for camera flow data, and the data rate requirements vary between 10 Kbps for vital signs transmission and 1.6 Gbps for 3D camera flow. Also, the BER ranges between 10^{-10} and 10^{-3}.

3.5 Disaster eHealth

eHealth can be effectively used in both man-made (war, pollution, fires, explosions, transportation accidents, etc) and natural (earthquakes, volcanoes, floods, fires, etc) disasters or infectious diseases to reduce their consequences (morbidity, mortality). Since disasters are unexpected and have shocking results, they require quick and effective intervention to preserve lives and properties and reduce suffering. This can be achieved using smart cards, wearable devices and mobile phones that may be helpful to collect, store and transfer medical data about the wounded [42]. Furthermore, Unmanned Aerial Vehicles (UAV) equipped with sensors and AI algorithms can be used to search and collect data of either victims or resources (number of victims, their location, available rescue resources, etc.) or to transfer small aid packages. According to Ref. [43], UAV applications require a high level of positioning accuracy (0,1–0,5 m), low latency ($<$ 150 ms), high availability (99–99,9%) and reliability.

For disaster eHealth, QoS requirements are low latency, low power devices, fast mobility, high reliability and secure communications.

4. Enabling technologies

In this section, we highlight how emerging technologies such as IoT, cloud computing, Edge/Fog computing, AI/ML and mobile 5G networks enhance healthcare services and satisfy their QoS requirements (Fig. 8).

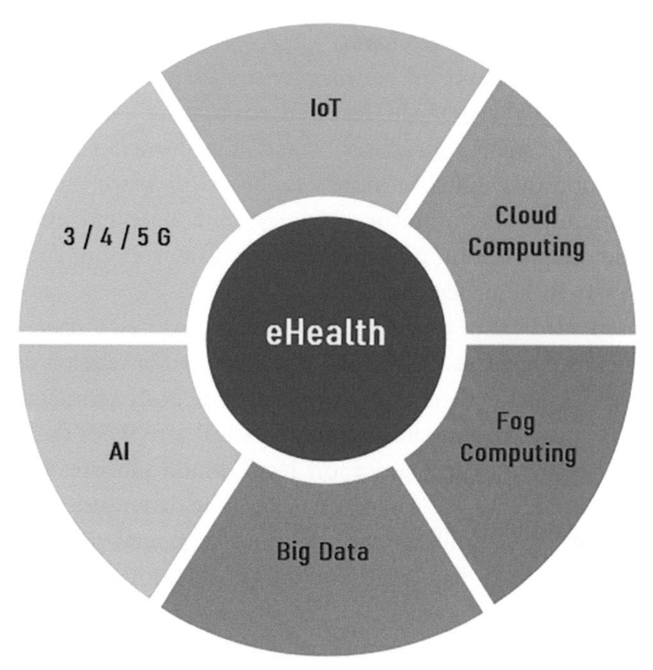

Fig. 8 Enabling technologies of eHealth.

4.1 Internet of things

The Internet of Things is a concept that defines the connection of everything, everywhere, and whenever to provide multi-sector services such as transport, smart cities, smart factories, healthcare, etc. It can be defined also as the convergence of information technologies, electronic sensors, cloud computing, network protocols, and Big Data via the Internet to provide specific services. In the healthcare context, IoT innovations and mobile applications help to drive major medical issues.

IoT eHealth solutions are nowadays widely adopted due to their ease of use, reduced cost compared to classical medical devices, accessibility and availability, and energy efficiency [44]. However, due to hardware limited capacity (CPU and memory), IoT devices alone are not able to process data, store it and guarantee the QoS of these services. So, to mitigate this issue, IoT-based eHealth solutions rely on cloud computing for storage and processing data and communication technologies to ensure traffic transport between IoT devices and cloud datacenters. Hence, the general architecture of an IoT-based eHealth system consists of three main layers: things layer, network layer, and core layer [15].

4.1.1 Things layer (or device layer)

It includes a large variety of sensors, wearables, and smartphones that collect patient's vital signs, activity, and environmental parameters such as smartwatches, smart eyeglasses, smart thermometers, smart shoes, ECG/EKG monitors, glucose monitors, blood pressure monitors, pulse oximeter, hemoglobin monitor, activity and sleeping monitor, etc. [14,44]. They require communication protocols to transmit collected data to upper layers to be processed.

IoT-based medical devices can be classified into two categories [15]: physical and virtual sensors. The former can be wired or wireless devices that measure the physical wellness of patients such as body temperature, heart rate, blood pressure, electrocardiogram (electrical activity of the heart), respiration rate, blood oxygen saturation level, blood glucose, body movement, hydration level, UV radiation, etc. The latter use software and mobile applications to collect vital and environmental parameters useful for remote monitoring, diagnostic, remote consultation, and patient health records.

Medical sensors are becoming more and more accurate, reliable, and low-cost. They can be widely distributed to different public hospitals, clinics, and patients at home in order to constantly track the health status of patients. However, these geospatially dispersed sensors produce enormous streams of sensing data which must be carefully processed and stored. Besides, an important challenge to overcome with IoT devices is the energetic limitation.

4.1.2 Network layer (or communication layer)

This layer provides communication technologies and protocols to ensure data transmission and processing between things layer and the core layer ranging from personal area networks to wide area networks [44]. Body and personal area networks that are commonly used are Bluetooth, ZigBee, Bluetooth Low Energy (BLE) and IPv6 Low power Wireless Personal Area Networks (6LoWPAN). Local area networks that can be exploited are Ethernet and Wi-Fi. Finally, cellular mobile networks (4G and beyond) are examples of wide area technologies that can be applied. In particular, 5G standard offers significant opportunities to support new healthcare services that will be highlighted later.

Recently, a family of technologies allowing long-distance communications with an extremely limited energy consumption, so-called Low-Power Wide-Area Networks LPWAN, has gained great attention thanks to their encouraging characteristics and promises in the IoT ecosystem compared to the traditional solutions (Wi-Fi, Zigbee, 6LoWPAN) [25]. Most of them are based on a cellular architecture in which end-devices communicate directly with base stations (or a central gateway). They offer a wide coverage reaching 10 km for urban scenarios and 20 km in rural scenarios. Besides, they are characterized by a low-power consumption comparable to PAN technologies such as Zigbee or 6LoWPAN. Furthermore, LPWANs architectures allow high density of user equipment's. These interesting features are attained thanks to the use of highly reliable modulation and coding techniques. The latter allow long transmission distances, with a very low-power transmission and high robustness to noise and interferences. This is of significant importance for medical IoT devices, as they are supposed to operate in a basis of 24/7 and in different indoor and outdoor scenario with minimal energy consumption. LoRaWAN, Sigfox, and Narrow-Band IoT (NB-IoT) are currently the most relevant LPWAN solutions.

4.1.3 Core layer (or cloud layer)

This layer provides [14,45]:

- Cloud gateways: They provide different protocols to establish the connection between IoT sensors and devices, and the cloud. The applied protocol depends on the technology used in the network layer.
- High-performance and scalable computing datacenters: They are needed to perform complex algorithms for data processing and huge long-term storage.
- Centralized management and control of users and devices: Since the cloud layer incorporates data from different users and devices, it is of vital importance to manage permissions and roles for users, IoT devices, and edge/fog nodes.
- Data management: The cloud layer is responsible for data collecting, automating, processing, long-term storage, and visualization. Therefore, the information can be accessed by all the stakeholders (patients, families, medical staff, insurance companies) at anytime, anywhere, and for any purpose (billing, visualization, etc.).
- Big data analysis: This is a central component of eHealth data processing. To make meaningful insights about patient's health, many algorithms can be applied to stored medical records such as data mining, ML, and automated reasoning-based algorithms. The results may be used to take preventive steps against deadly diseases or possible secondary effects. For example, in Ref. [46] ML models were used to identify

and classify different types of brain tumor in early stages. In Ref. [47], the authors present a ML-based approach to diagnose diabetes from EHRs. And Ref. [32] defines a prototype that focuses on the early detection of cardiopathies.

4.1.4 Challenges and new trends in internet of medical things

IoT features offer big opportunities for eHealth to support data collection from controlled patients thanks to their ease of use, small size, and reliability. However, they suffer from limited computational capabilities, energy consumption and security and privacy issues.

Indeed, although most of wearable devices include a new generation of performant microcontrollers as processing units, their computational power is still far from modern CPUs. Hence, it is difficult to execute some heavy computing tasks in the end devices. A new trend is to offload those tasks to be executed in Edge/Fog devices.

The available energy is an important factor to consider and optimize for IoT-based eHealth solutions. In fact, medical sensors are supposed to work on a 24/7 basis and any failure can endanger patient's life. Hence, it is necessary to avoid frequent battery charging or replacement as this is cumbersome and even impossible for many cases such as medical implants. Energy consumption is due to processing tasks and data transmission through wireless networks. Although the use of the new generation of processing algorithms (such as TinyML that will be seen later) and low-power networks (LPWAN) allows the reduction of power consumption, the small size of batteries is yet challenging. Recently, a promising field of research has gained a lot of attention which is energy harvesting. It aims at collecting and converting small amounts of energy from surrounding sources such as light, vibrations, radio waves, solar energy, the device carrier's movement, etc. The collected energy is then used to power the wireless devices making them entirely independent from the power grid [25].

Security and privacy are another two-related issues to consider in IoT-based eHealth schemes. The latter are vulnerable to many types of threats due to the following facts:

- eHealth devices and sensors, which manipulate and exchange personal data, are mobile and the user may connect to different networks when moving. So, they are exposed to attacks such as the interruption of the service, modification of the transmitted data, replaying the messages, etc. Traditional cryptographic algorithms cannot be applied because of the limited computational and memory features of end devices.

- Medical devices are connected to each other or to the cloud through different types of network infrastructures and protocols involving wireless, wired, private, and public infrastructures (Bluetooth, Zigbee, WiFi, Ethernet, cellular networks, etc.) which makes the end-to-end communications vulnerable to several security issues like data interception, integrity violation, inappropriate QoS, etc.
- eHealth applications run on the end terminals (smartphones, tablets…) involve web applications, middlewares, interfaces, etc., so they are vulnerable to traditional software attacks such as stealing data, web parameter tampering, bypassing authentication or compromising database integrity [48].

ML methods coupled with Edge computing are gaining more attention to implement security procedures (mutual authentication, access control, etc.) in the edge network to cope with the limited computational power of end devices.

4.2 Cloud computing

According to the National Institute of Standards and Technology (NIST) special publication [49]: "cloud computing is a model for enabling ubiquitous, convenient, on-demand network access to a shared pool of configurable computing resources (e.g., networks, servers, storage, applications, and services) that can be rapidly provisioned and released with minimal management effort or service provider interaction".

In the context of eHealth, cloud computing can be used to facilitate communication, collaboration, and coordination between different stakeholders which reduces costs and increases the effectiveness of healthcare services.

Cloud computing offers the following advantages to eHealth services [11]:

- Patient electronic records are hosted on the cloud. So, they are available for different stakeholders for analysis and diagnosis at anytime and anywhere.
- With cloud services, there is no need to buy its own servers and applications neither to support maintenance costs. They are provided, managed, and maintained by cloud providers. Their deployment is more rapid and flexible.
- Doctors in remote areas can use cloud resources to perform remote consultations.

- In the case of disasters, data recovery is easier since cloud providers offer redundant systems.
- Researchers can benefit from central medical data hosted on the cloud to conduct their research, to control diseases, and to monitor epidemics.

However, the centric scheme of eHealth data and applications has many limitations related to cloud constraints [50]:

- The service quality is not always guaranteed. It depends on internet connection characteristics, location, technologies, and circumstances. As eHealth services must be available continuously with the same performance, cloud resources must be provided and managed to counter any interruption and degradation of service.
- Interoperability between different cloud platforms is not yet supported. Hence there is a serious need to define standards to ensure interconnexion and collaboration between eHealth provider's platforms.
- Cloud hosted data are more vulnerable to data loss, theft, and malicious attacks. Cloud security in one of the most challenging issues in eHealth industry. It is related to authentication and authorization procedures, identity theft, data encryption, etc.

4.2.1 Challenges of cloud-based eHealth

Cloud computing is considered as a key enabler for promoting the healthcare industry. It offers many advantages like flexibility, fast deployment, cost and energy effectiveness, collaborative resource sharing, etc. However, it still raises many issues mainly related to latency, security, and privacy.

According to the 3-layers architecture of IoT, end devices rely on the cloud layer to perform processing and storage tasks. However, in the case of time-critical healthcare solutions, excessive exchange of data between these two layers may produce extra latency. A new layer, called Edge layer, was defined to execute traditional cloud tasks (processing, mid-term storage, security procedures, etc.), thus enhancing the end-to-end latency [19].

As for security issues, EHRs contains both personal and medical sensitive data that are transmitted and stored through shared networks and servers. They are exposed to many threats such as tampering, unauthorized access, denial of service attacks, spoofing identity, etc. Several approaches have been proposed to preserve security and privacy of health data in the cloud. Cryptographic schemes as well as fine-grained access control guarantee data confidentiality and integrity and enforces patient-centric data access. Recently, blockchain technology was proposed by many research works to guarantee data integrity, fine-grained access control, traceability and transparency [51].

4.3 Edge/fog computing

There have been many research efforts to mitigate some of the limitations of cloud-based eHealth services by moving the computational power closer to the end devices, in the edge of the network, to support delay-sensitive applications especially when computing results are needed by the data generators. Many paradigms have appeared such as Edge computing, Fog computing and cloudlet. Although they propose different implementations, but they are based on the same concept. In the following, Fog and Edge computing will be used interchangeably.

Fog computing is a distributed computing concept derived from cloud computing. Introduced by Cisco, fog computing is defined as "a highly virtualized platform that provides compute, storage, and networking services between end devices and traditional cloud computing datacenters, typically but not exclusively located at the edge of network." [52].

It is based on Fog Nodes (FN) that can be deployed everywhere near IoT devices. The function of a FN can be performed by any device with computing, storage, and network connectivity (switches, routers, servers, video surveillance cameras, etc.) [52].

4.3.1 Edge/fog computing achievements in eHealth

As stated in the above section, even though the cloud offers many advantages for IoT-based eHealth systems related to communication, data storage and processing; it fails in addressing some issues [15]. In fact, in real time applications, transferring data to the cloud to be processed and delivering the result back to the IoT devices produces extra delays and additional overheads. Additionally, situations leading to application unavailability due to any cause (power failure, loss of internet access, cloud failure, etc.) can be life-threatening.

To rectify the above-mentioned problems, many research works introduce an additional Edge/Fog layer between IoT devices and the cloud core [14]. Health state data are generated by IoT end-devices and forwarded to the fog layer. The latter is composed of multiple Fog Nodes. Each one serves a local group of health devices and performs delay-sensitive computing tasks to make rapid decisions, the other traffic is relayed to the main storage devices in the cloud. The fog layer reduces the burden by executing network functions such as switching, routing, and network analysis. Additionally, they perform data processing such as compressing, filtering, aggregation, and formatting. Finally, they can carry out some security measurements such as access control, authentication, and encryption.

The network bandwidth is another critical issue in eHealth applications resulting from huge data transmission. The authors of [15,14] deal with the example of ECG (electrocardiogram) devices that generate gigabytes of data within a day. For thousands of patients, it is not feasible to send this large amount of data to the cloud. In Ref. [15], a collaborative ML approach is proposed where the intelligence is distributed between device layer, fog layer and cloud layer. The given architecture revealed better performance in terms of latency and energy consumption.

In Ref. [53], the authors study the effectiveness of the fog layer on energy saving and delay optimization by defining an energy-efficient strategy that allocates incoming tasks to fog devices based on the remaining CPU capacity and energy consumption. Through Remote Patient Monitoring (RPM) as a case study, the results show that using the fog layer reduces the consumed energy and minimizes the application round-trip delay. In Ref. [28], a system for monitoring elderly health based on IoT and Fog computing is proposed: physiological parameters (body temperature, heartbeat rate, etc.) are measured by a wearable system and sent to an Android application. The latter performs the role of a fog server. It analyses the received data in real time, the general parameters are stored locally, and permanent data are sent to the cloud for the permanent storage to be available for the patient, his family, or his caregivers. In Ref. [29], the authors test the "eWALL" which is a home monitoring system composed of a passive infrared movement detector and a gas sensor to detect the presence of smoke in the breathed air. The fog nodes provide real-time notification if the user falls down or if the oxygen level is out of the normal range; while the cloud is requested to store the patient history or to retrieve it if needed. Another remote monitoring eHealth system was evaluated in Ref. [54]. It consists in collecting both patient's signs (heartbeat rate, body temperature, and electrocardiogram) and environmental parameters of the room (light and noise level, indoor and outdoor temperature, and humidity) and sending them to a fog server located on a Raspberry Pi Zero W board from Adafruit. In case of any abnormal value, the system alerts the remote patient's caregivers. Mobility and security are other important issues that was considered in Ref. [55] via a smart gateway located in the fog layer that performs local data processing for real-time notification, authenticating sensor nodes and supporting mobility of patients to prevent data loss and service disruption.

4.3.2 Open issues of fog computing in eHealth

Fog computing was considered by recent research works as a promising approach to support delay-sensitive applications and efficient-energy consumption for eHealth end-devices and minimize data traffic in the core network. However, there are some constraints to be considered in future research [54] which are mainly:

- Complexity: A FN processes and analyzes its own set of data independently. Hence, there is no guarantee that the resources from a specific node will be the same in the other ones which adds more complexity to the network especially for the management and maintenance operations.
- Power consumption: Introducing FNs in the network architecture needs extra power supply for them to operate.
- Security: FNs are usually held by different owners. This fact makes it difficult to guarantee the minimum level of privacy and protection. A particular Fog Node can introduce malicious software which may affect the overall system or makes it work incorrectly.

So, a trade-off between complexity, efficiency and security must be considered in the fog layer.

4.4 Machine learning/deep learning

ML is a branch of AI where systems can learn from collected data to identify and build analytical models/patterns without being explicitly programmed. The latter are used to make decisions with minimal human intervention. A typical ML algorithm is made of two phases: training phase and testing (or inference) phase. Firstly, inputted data are used to make a prediction or classification model, which is called training. Then, the model is applied on a new dataset to test and validate the model. DL is a sub-branch of ML that imitates the working of human brain to establish the model.

4.4.1 ML/DL applications in eHealth

eHealth applications generate an enormous and heterogenous amount of data daily collected from multiple sources (medical sensors, medical images, environmental data, etc.). ML/DL methods help to extract meaningful information and process them more effectively than traditional statistical models. ML/DL algorithms were widely used in eHealth for prevention, diagnosis, real-time remote monitoring, EHRs data processing, etc. Different schemes were proposed in the literature ranging from executing ML methods on end devices (on-device ML), on the edge (Edge ML), on the cloud or jointly in multiple layers as shown in Table 3. In Ref. [56], an on-device prototype called SPHERE (Sensor Platform for Healthcare in a Residential Environment) was proposed for long-term

Table 3 ML/DL applications in eHealth.

Refs.	Application	ML/DL algorithm	Challenges	End devices
[57]	Recognition and classification of skin cancer	CNN	Accuracy of image classification	Smartphone
[58]	Fall detection for elderly	BDT	–	Accelerometer, gyroscope
[59]	Prediction of pregnancy-related risk situations	AODE	Computational complexity	–
[60]	ECG arrhythmia classification	CNN/ SVM	Computational complexity	Raspberry Pi
[61]	EEG-based seizure detection, ECG-based arrhythmia detection	SVM	Energy consumption	EEG/ECG sensors
[62]	ECG and activity classification	RF/SVM	Classification accuracy	ECG/ activity sensors
[63]	Activity classification	SVM, Decision tree, Gaussian naive Bayes (GNB)	Energy consumption, Latency	Accelerometer, ECG/temperature/ humidity sensors, Raspberry-Pi 3
[64]	Activity classification	SVM	Battery Lifetime	SPHERE
[65]	Fall and accident detection for elderly patients	DNN	Inference process acceleration	Nvidia Jetson NX platform
[66]	Brain tumor segmentation	DNN	Tradeoff between privacy-accuracy	NVIDIA Tesla V100 GPU

residential monitoring and human activity classification. Signals are collected by an accelerometer. Then, the Support Vector Machine (SVM) algorithm is applied to classify human activity into three categories: sedentary, moderate, and sportive. Finally, data are sent through a BLE radio to a central unit. In Ref. [57], Dai et al. present an on-device application for skin cancer detection based on the Convolutional Neural Network (CNN). The latter is applied by a smartphone to classify images of skin lesions. The authors of Ref. [58], evaluated the accuracy of four ML algorithms (Random Forest (RF), artificial Neural Network (ANN), SVM and Boosted Decision Tree (BDT)) for fall detection in elderly monitoring application. The proposed system called Whoops is made of a smartphone connected to an accelerometer, and a gyroscope is used for sensing, collecting, processing, and sending data to the datacenter. The algorithms were implemented in the Edge network then in the Cloud. Cloud-based solution gave better performance regarding accuracy. However, Edge-based solution reduces the needed bandwidth and storage memory. The work in Ref. [59] applied the Averaged One-Dependence Estimator (AODE) to identify high-risk situations for pregnant women with hypertensive disorders of pregnancy that may lead to serious problems for both the mother and the fetus. Based on a set of attributes such as symptoms, risk factors, and physiological indicators; the AODE classifier can predict different kinds of childbirth delivery for pregnant women who suffered from hypertension during their pregnancy.

4.4.2 Challenges and new trends of ML/DL in eHealth

ML/DL are greatly enhancing eHealth services. However, there are still numerous issues regarding algorithm complexity, energy consumption, latency, and security. Indeed, IoT devices are resource-constrained (memory, processing power, energy) while traditional ML algorithms such as image processing consume a lot of processing power and memory. This can be addressed by offloading some tasks to the edge or cloud layer. However, cross-layer exchanging of medical data may affect QoS requirements of real-time applications, mainly bandwidth and end-to-end latency. In addition to QoS issues, inter-layer ML/DL execution exposes private and vital personal data to many security threats such as data interception, deletion, or modification.

Many approaches to guarantee QoS optimization along with ML accuracy were proposed in the literature which resulted in new emerging paradigms such as TinyML and Federated Learning. TinyML is a new research field that includes hardware and software solutions aiming at running ML analytics at lower-constrained edge devices. Typically, the power consumption for such systems is in the mW range [67]. Several robust hardware and software frameworks executing both training and inference phases in the edge have been proposed. To name a few examples, Nvidia Jetson

NX and Google Coral [68] are two small low-power hardware toolkits that were designed to execute on-device high-speed AI inferencing for applications like image classification and object detection. TensorFlow Lite and PyTorch are examples of software tools designed to deploy ML algorithms on tiny and low-power end devices [67]. For example, in reference [65], the authors develop an on-device Healthcare Body-pose Estimation App named as Kestrel that monitors elderly patients to detects abnormal movements (fall off their bed, accident, etc.). The application run two existing DNN-based object detection algorithms (RetinaFace and OpenPifPaf) on the Nvidia Jetson NX platform (contains GPUs, DL hardware accelerators) integrating the TensorRT library. If the monitored patient falls off his safe position, an alarm is raised. The proposed system has shown good performance regarding fall detection accuracy. However, the inference process time needs more acceleration.

Federated Learning (FL) is another emerging technique that has been recently proposed by Google Inc. to mitigate the problems of bandwidth loss, latency, and data privacy [69]. An algorithm is trained across distributed edge devices using their own local data. Then, the local weights returned by each device are shared with a central server which collects the local weights and computes the new global parameters. Hence, FL guarantees the privacy of local data during the training stage since they are not shared among the collaborating parts. FL was applied in Ref. [66] to model brain tumor segmentation. The proposed system is based on a client-server architecture. A centralized server executes a global DNN model (on a NVIDIA Tesla V100 GPU) and collects coordinating clients' local parameters updates. The client-side is designed to have full control over its local data. Performance evaluation regarding accuracy and privacy preservation was conducted for different shared proportions of local data (of the clients). According to the experimental results, sharing larger fractions of models gave better performance.

To conclude with, edge intelligence, QoS provision (latency and energy) for medical use cases and security/privacy requirements are three interdependent parameters that must be treated carefully.

4.5 5G mobile networks

The fifth generation (5G) of cellular networks was defined to significantly improve the offered QoS over the preceding generations, specifically in terms of higher data rates, lower latency, ultra-high reliability, higher connection density, and higher mobility. Indeed, compared to the fourth generation, it was envisioned that data rates will be improved by 10 to 100 times, as well as a higher traffic density around 1000 times reaching a density $\geq$ 1 M terminals/km^2 and a decreased end-to-end latency by a factor of 1/5 [8]. The International Telecommunication Union (ITU) defined three generic traffic classes for 5G cellular networks [70]: enhanced Mobile BroadBand (eMBB), ultra-Reliable Low Latency Communication (uRLLC) and massive Machine Type Communication (mMTC). Each class has its own QoS Key Performance Indicators (KPIs):

- eMBB: This category focuses on signals with high data rates, low latency, and high connection density. In the eHealth context, it can be useful for transmitting very high-definition video signals in the case of remote surgeries and communications between 5G autonomous ambulances and hospitals [71].
- uRLLC: This category was defined to handle ultra-reliable low latency communications. Such requirements are needed for autonomous ambulances, tactile Internet applications and remote robotics to perform remote surgeries. Mobility, latency, and reliability are the major requirements for such use case.
- mMTC: This class is designed to support a big number of connected devices with low bandwidth transmissions (small packets and sporadic traffic) and low energy consumption. It is the most suitable class for IoT devices without critical time constraints such as connected fitness bands, smart watches, etc. Network energy efficiency and connection density are the main requirements in this case.

4.5.1 5G QoS indicators

In this section, we describe the key performance indicators that have been defined by the ITU to specify, quantify, and measure the characteristics of 5G systems [72]. Then, we will use these KPIs to characterize eHealth services. Table 4 defines the eight basic indicators and gives their associated traffic category.

Evaluation of 5G services is not limited to those 8 KPIs, there are other metrics that may be used to make usage scenarios more flexible, reliable, and secure such as [72]:

- Spectrum and bandwidth flexibility: refers to the flexibility of the system to support different scenarios for different frequency ranges.
- Reliability: refers to the maximum tolerable packet loss rate at the application layer.

Table 4 ITU key performance indicators for 5G networks.

KPI	Description	Use case
Peak data rate (in Gbit/s)	Maximum attainable data rate per user/ device under perfect conditions.	eMBB
User experienced data rate (in Mbit/s or Gbit/s)	Reachable data rate that is accessible ubiquitously to a mobile user/device in the coverage area.	eMBB
Latency (in ms)	Maximum acceptable end-to-end latency from the time a data packet is generated at the source application to the time it is received by the destination application.	uRLLC
Mobility (in km/h)	The maximum speed at which it is possible to achieve a given QoS.	uRLLC
Connection density	Total number of connected devices per km^2.	mMTC
Energy efficiency (in bit/J)	Quantity of information per unit of consumed energy.	mMTC
Spectrum efficiency (bit/s/Hz)	Average data rate per spectrum resource unit and per cell.	eMBB
Area traffic capacity (in $Mbit/s/m^2$)	Traffic throughput served per geographic covered area.	eMBB

- Resilience: it is the capability of the network to operate appropriately even in the case of natural or man-made disturbance, such as the loss of main energy supply.
- Security and privacy: refers to providing adequate protection services for users, data and network resources against attacks.
- Operational lifetime: refers to the maximum operation time performed per stored energy capacity. This indicator is mainly significant in the case of IoT devices requiring a very long battery life because it is too hard to perform regular maintenance due to physical or economic reasons.
- Positioning accuracy: refers to the maximum positioning error tolerated by the application.

4.5.2 Matching 5G key performance indicators with eHealth requirements

Since the initial stages of 5G standardization in 2015, eHealth applications have been identified as one of the most relevant and challenging verticals [8]. Theoretically, 5G promises can perfectly meet the strict QoS requirements of eHealth use cases mainly related to latency, user experienced data rate, reliability, connection density, and mobility.

These promises have been proven through a series of 5G trials and commercial deployments presented in Ref. [23]. The discussed experimental performance reports demonstrated that theoretical healthcare requirements are indeed possible to meet with existing 5G capabilities (see reference [23] for more details). However, the authors note that most of the presented experimental reports describe individual KPIs in detail but trade-offs between multiple KPIs in healthcare applications are rarely considered. For example, HD video transmission needs a trade-off between throughput and latency. Similarly, massive IoT-based monitoring in a dense urban area requires a trade-off between coverage, capacity, and load balancing.

For each eHealth scenario, signals are mapped into a traffic category (eMBB, uRLLC, mMTC) depending on their QoS requirements in order to be assigned a set of network resources. However, the scenarios described above may fall at the intersection between these categories. Fig. 9 illustrates some medical use cases with their corresponding 5G-defined categories.

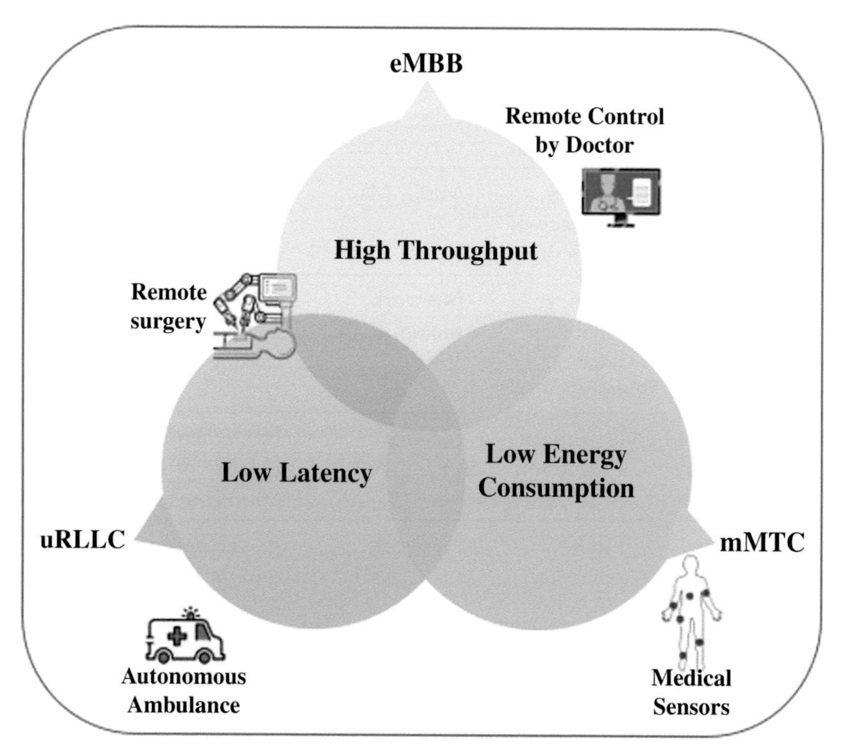

Fig. 9 Examples of eHealth use cases mapped to 5G-defined categories.

For example, medical probes have the characteristics of both mMTC and eMBB [71]: since probes transmit low power signals, they are classified as mMTC, but if they transfer ultrasound scan, they require high bandwidth transmission provided by eMBB. Remote surgeries require real-time high-definition video transmission with ultra-low latency, which means they have the characteristics of both eMBB and uRLLC. Remote consultations are based on high-definition video streaming, so they are categorized as eMBB traffic.

The "Connected Ambulance" as defined by the 3GPP supports real-time streaming of the handled patient parameters to the awaiting emergency crew at the destination hospital. It is considered as an eMBB since it requires both extremely high bandwidth transmission, high mobility, and reliable broadband access over large coverage areas [73]. Finally, autonomous ambulance needs ultra-reliable communications, so it belongs to the uRLLC category. In Table 5, we collected some research works that attempt to implement different use cases through 5G networks.

Table 5 eHealth scenarios through 5G networks with their requirements and realizations.

Refs.	Use case	QoS requirements	Research topics	Realization
[41]	Telesurgery through 5G	Latency, high throughput	Edge-cloud, network slicing	An architecture based on 5G tactile internet
[32]	Remote monitoring through 5G	Latency, accuracy of prediction	MEC, ML	An architecture and protocol for smart monitoring
[73]	Connected ambulance	Latency, reliability, large coverage areas	Network slicing, MEC, VNF	VNF deployment in the edge to execute low-latency processing
[74]	Connected ambulance	Quality of Perception (QoP) metrics	Network slicing, ML	Implementation of a Quality of Experience optimizer based on VNFs
[27]	Remote monitoring through 5G	High-reliability, efficiency	Network slicing, ML	A web application for remote monitoring
[75]	Remote monitoring and ML prediction	Computation efficiency, latency	MEC, NFV, network slicing, orchestration, ML	An end-to-end 5G service orchestration platform for remote health monitoring and prediction

In Ref. [8], the 5G Public-Private Partnership (5G-PPP) working group give an exhaustive study about different eHealth scenarios and categorize them into four main use cases. Table 6 summarizes the QoS requirements for each one.

4.5.3 Technology trends in 5G networks to satisfy eHealth requirements

One of the most challenging features of eHealth traffic is its heterogeneity and conflicting requirements in many scenarios. Such is the case of telesurgery or connected ambulance where extremely low latency and high data rate video traffic could coexist with much lower data rate but massive traffic issuing from wearable IoT devices. To meet these requirements, 5G must implement intelligent techniques by setting priorities and aggregating data through many techniques such as network slicing and edge-AI-based techniques [22].

4.5.4 5G slicing

Network slicing is a process that creates multiple independent, logical, and virtualized networks over a common physical infrastructure. The purpose is to provide logical network resources to satisfy specific application demands. Physical components and resources may be shared throughout network slices, but each one has its own architecture, management policy and security. QoS metrics over each slice are provided to comply with a particular Service Level Agreement (SLA).

Network slicing is considered as the most remarkable shift between the fourth and the fifth mobile network generations. It is based on a set of functions: Software-Defined Networking (SDN), Network Function Virtualization (NFV), orchestration, etc. SDN is an approach that brings intelligence and dynamic programmable network functions by separating the control plane from the hardware of the network. External data control is performed through a logical software entity called a controller located between network devices and applications.

NFV is a 5G specific concept that uses virtualization to enable network functions such as network forwarding, firewalling, encryption, etc. Thus, network functions are moved from dedicated hardware to virtual servers. NFV is a very hot discussion topic that many researchers have worked on specially to define optimization algorithms on how to decompose Network Services (NSs). Orchestration consists in interpreting and mapping service QoS requirements into configuration of physical and virtual resources needed for service deployment [76].

Table 6 5G-PPP defined eHealth use cases with their corresponding QoS requirements.

Family of use cases	Examples	QoS requirements	Involved KPIs
Assets and interventions management in Hospitals	– Assets tracking and management (wheelchairs, ECG monitors, infusion pumps, etc.) – End-to-end monitoring and management of pharmaceuticals through the supply chain. – Intervention planning and follow up (smart management of waiting lists for surgeries, data on availability of the clinical staff, patient's clinical parameters, etc.)	– Scalability in terms of the number of devices connected: for each hospital, thousands of linked objects need to be monitored. – Positioning accuracy: must not exceed 1 m in indoor conditions. – Seamless handover to guarantee service continuity between multi access technologies. – Mobile assets must be supported (for example medical helicopters): mobility >300 km/h. – Strong security mechanisms	– Connection density – Positioning accuracy – Mobility – Security and privacy
Robotics	– Robotics assisted tele-surgery. – Robotics assisted living	– Ultra-high reliability for data transmission: connectivity should be maintained in all circumstances (natural disasters). – Low latency: in the order of 30 ms – Strong security mechanisms	– Reliability – Latency – Resilience – Security and privacy
Remote monitoring of health or wellness data	– Delivering medical care at home through wearable monitoring systems especially for patients having chronic diseases – Lifestyle and prevention. – Follow up after severe incidents to optimize hospital resources and reduce healthcare costs	– Improve coverage: guarantee continued coverage in both rural and urban areas. – Improving the consumption of energy and battery life of connected devices (wearables/implants): the goal is to support energy supply for at least 10 years without battery replacement. – Mobility at high speeds (>300 km/h) – Strong security mechanisms	– Connection density – Energy efficiency – Resilience – Operational lifetime – Mobility – Security and privacy
Smarter medication	– Applying medication to the patient remotely if needed to prevent dramatic effects and raise recovery chances (from brain stroke for example) – Smart pharmaceuticals powered by connected devices	– Better coverage and mobility – Improving energy efficiency – Scalability in terms of the number of connected devices – Strong security mechanisms	– Connection density – Energy efficiency – Mobility – Operational lifetime – Security and privacy

Each defined traffic category (eMBB, mMTC, uRLLC) corresponds to a network slice. So, every new service is mapped into a traffic category and handled through the corresponding slice. A 5G operator should manage the creation, modification, and deletion of a slice without any impact on traffic and services in other network slices.

Handling eHealth traffic through 5G slices was discussed in many research papers and many prototypes of eHealth systems were developed and tested. SliceNet, which is a 5G-PPP project that focuses on network slicing in 5G networks [39], proposed a 5-layered model for 5G slicing and tested the proposed architecture for different systems in different verticals. As for eHealth vertical, researchers focus on the "Connected Ambulance" use case. In the proposed scenario, the connected ambulance acts as a connection hub for the emergency medical devices and wearables and transmits high-definition video to the destination hospital. Hence, it is an eMBB service, requiring extremely high data rates, low-latency, and reliable communication over large coverage areas [77].

In Ref. [40,77], the authors attempt to validate the benefits of the SliceNet architecture in the "Connected Ambulance" use case. They focus on QoS guarantee for real-time video streaming from the ambulance to the hospital by introducing an edge algorithm to reduce the latency. It is based on a ML analysis to detect possible signs of stroke and sends the results to the hospital. In Ref. [27] proposes an eHealth system called "Dr. Pocket" for collecting and processing medical data from heterogenous IoT devices connected via 5G network. The satisfaction of the different requirements of the connected devices is performed through dynamic provision of 5G slices. Data processing is realized in 4 steps. The first one is data collection by a discrete Network Service (NS) where medical data are gathered from IoT devices and translated into service requirements to identify if there are some conflicting requirements. The conflicts are analyzed and resolved. The second step is network slices placement: an end-to-end slice is allocated based on the application's requirements, the network availability, and the cost. The third stage is data analysis which consists in comparing the collected data to previous analyzed data using a ML approach (K-means clustering algorithm). Finally, collected data are transformed into comprehensive information to future decisions. In Ref. [78], the authors introduce NFV service federation and give a demonstration for an eHealth Emergency Service use case deployment. NFV service federation is a concept that was defined by the 5G-TRANSFORMER (5GT) project [79] aiming to deal with multi-provider deployment and management of vertical services.

The tests mentioned above showed that network slicing is a powerful 5G feature but in most cases, it must be powered by Edge computing and ML.

4.5.5 Multi-access edge computing (MEC)

MEC refers to deploying Edge computing tasks (such as data caching, video analytics, augmented reality, location services, etc.) in cellular networks, specifically in the Radio Access Network (RAN) or macro–Base Station (BS), this provides higher bandwidth, lower latency, and context and location awareness [80].

Recently, abundant research work have focused on the contribution of MEC in 5G based-eHealth systems, particularly when coupled with AI or ML. In Ref. [41], the authors introduce a telesurgery robot based on a 5G tactile network. In the ideal network conditions characterized by a maximum end-to-end latency of 100ms and an ultra-low packet loss rate, the robot receives commands from the distant doctor to execute the corresponding procedure. In the case of overtime, an AI algorithm is executed on the edge to predict the robot's next task to execute. Ref. [32] proposes a smart eHealth architecture and protocol for remote monitoring (ECG, heart rate, activity) of chronic patients running different AI algorithms (Neural Networks, rule-based system) to detect arrhythmia pathologies from the collected parameters. Silva et al. propose in Ref. [81] a Local 5G Operator (L5GO) architecture for delay critical telehealth services located in the edge network. They consider two use cases: AR-assisted and robotic-aided surgery. The simulation results showed that the performance of the proposed architecture in terms of latency is better than traditional architecture. In Ref. [73], virtual functions were deployed in the edge to enhance the latency and the coverage in the case of the connected ambulance. In Ref. [34] proposes a Real-time Heart Attack Mobile Detection Service (RHAMDS) which takes advantages from SDN and MEC VANET (Vehicular Adhoc Network) features to reduce the response time of emergency aid for heart attack patients moving on vehicles and avoid possible collisions.

To conclude with, most recent researches agree [18,75] that 5G-based eHealth services such as remote monitoring, remote consultation, remote surgery requiring ultra-high-definition video quality and massive data transmission cannot be achieved without a smart Edge layer running IA/ML algorithms to enhance QoS provision. Nevertheless, Edge computing raises two main challenges. Firstly, the complexity of computational tasks is proportional to energy consumption, so a trade-off between effectiveness and energy consumption must be considered. Second, because of the distributed location of Edge devices, they are vulnerable to many types of security attacks such as DoS, man-in-the-middle, data tampering, privacy disclosure, etc.

4.5.6 Challenges of 5G mobile networks in eHealth

eHealth was considered by the 5G mobile networks as a separate vertical characterized by its own QoS requirements. 5G slicing is a powerful characteristic that helps in provisioning QoS mainly in mobility management, high coverage, low latency, high data rates, and extra high reliability. Furthermore, the combination of 5G, Edge computing and ML has paved the way for the advancement of "smart medicine." However, 5G-based eHealth still faces a lot of challenges related to the following facts:

- Slicing mechanisms are yet in their infancy, they need much more work principally in NFV and orchestration.
- MEC enhances considerably QoS provision for eHealth services mainly for real time and latency-constrained applications. However, it is still challenging in terms of security, energy efficiency and QoS satisfaction.
- Security is also an important issue to focus on in the different layers of 5G networks: virtualization security, authentication schemes, data integrity, privacy protection, service availability, identity threats, location privacy, etc.

5. Open issues and challenges of eHealth

As discussed below, new technologies such as IoT, Cloud/Fog/Edge computing, AI/ML and 5G mobile networks have led to significant improvements in the healthcare industry. However, there are still many research issues and challenges to focus on to improve QoS and security requirements, mainly in the following areas, see Table 7.

- Data management and analysis: Medical devices produce an enormous volume of data that need rapid, smart, and secure transmission, processing, and storage. Therefore, intelligent algorithms and approaches are needed for data analysis in the Cloud/Fog/Edge layers to improve health monitoring, disease diagnosis and reduce time wastage. Such intelligent procedures imply many consequences:
  - Computation power for data processing must be provided via hardware and software optimizing tools.
  - Security and privacy must be provided for private and medical sensitive data.

Table 7 Summary of the contributions and challenges of emerging technologies on eHealth.

Technology	Opportunities	Challenges
IoT	– Remote monitoring – Ease of use – Cost reduction – Energy Efficiency	– Low computational and storage capacity – Security/data privacy
Cloud computing	– Effective, rapid, and smart (AI/ML algorithms) data processing – Long term storage of health records	– Extra latency produced by data transfer between IoT devices and the cloud – Security/data privacy
Fog/Edge Computing	– Reduced latency for real time applications – Real-time notifications – Short-term storage and processing – Security measurements	– Complexity of fog nodes – Energy consumption – Security/data privacy
IA/ML	– Smart data processing – Early diagnosis/prognosis	– Processing power – Energy consumption – Security/data privacy
5G Mobile networks	– Mobility management of patients and devices (health sensors, ambulances, etc.) – Positioning accuracy – Low latency – Extra high reliability – High bandwidth for high-definition video transport – High coverage	– Network slicing optimization – Mobility management – Security/data privacy

  – Energy consumption must be optimized for both computational and transmission procedures, a new promising generation of embedded powerful processors and low power networks has appeared (LPWAN) that can be integrated successfully in eHealth solutions.

- Edge optimizing capabilities: Processing data at the Edge network can reduce the latency, enable real-time notifications, optimize the bandwidth, and ensure the protection of sensitive data transmitted between patient and healthcare provider. Edge computing may be powered by AI/ML algorithms to speed up data processing. New trends are emerging for Edge intelligence such as TinyML and federated ML that aim at executing adapted ML models on end devices to reduce latency and enhance data security. However, there still some constraints regarding algorithms' accuracy and energy consumption.

- QoS provision: eHealth QoS metrics are very strict mainly for ultra-reliable real-time services. Latency is considered as the most challenging requirement as it is directly related to human life. Although all the above discussed techniques (Fog/Edge computing, AI/ML, 5G network slicing) improve the end-to-end latency, there still exist much work to do for real-time scenarios.
- Security and privacy: Since medical data are strictly related to personal privacy and security. Their protection is expected during their whole life cycle (collection, processing, storage and destruction) to enhance the trust level between all the stakeholders (patients, healthcare providers, insurance companies, etc.). In particular,
  - Authentication and authorization are extremely needed to guarantee fine-gained access control and prevent unauthorized access to private data.
  - Cryptographic schemes must be implemented with respect to device computational power and QoS constraints.
- Connectivity and coverage: eHealth data should be transferred safely, securely, at the right time, and in all circumstances (emergency, disaster, etc.) between geographically distant interferers. So, there is a need to guarantee seamless handover and interconnection between different networks and providers.

As a conclusion, an eHealth scenario is often a multi-dimensional problem where multiple metrics need to be optimized simultaneously.

6. Conclusion

This article focused on enabling technologies that have boosted the quality of healthcare services. The survey is divided into four parts. The first part described the typical architecture and main components of eHealth systems. The second part highlighted the QoS requirements needed for each use case. The third part has gone into the key features, challenges, and new trends for the enabling technologies of eHealth. Concretely, we focused on IoT, Cloud computing, Edge/Fog computing, ML/DL and 5G networks. Despite the advantages offered by these emerging technologies for eHealth, many challenges are yet to deal with mainly data management, QoS provision and security. In the future, security and privacy in eHealth will be highlighted in depth.

Declarations

Conflicts of interest

The Authors declare that there is no conflict of interest associated with this publication.

Funding

The authors did not receive support from any organization for the submitted work.

References

[1] J. Hathaliya, S. Tanwar, An exhaustive survey on security and privacy issues in healthcare 4.0, Comput. Commun. 153 (2020) 311–335.
[2] J. Corrigan, E. Swift, M. Hurtado, Envisioning the National Health Care Quality Report, National Academies Press, Washington, 2001.
[3] WHO, Global Diffusion of eHealth: Making Universal Health Coverage Achievable, Report of the third global survey on eHealth, ISBN 978-92-4-151178-0, Global Observatory for eHealth, 2016.
[4] WHO, mHealth: New Horizons for Health through Mobile Technologies: Second Global Survey on eHealth, World Health Organization, 2011.
[5] NIH, Mobile Health: Technology and Outcomes in Low- and Middle-Income Countries (R21), PAR-14-028, 2016.
[6] WHO, Telemedicine: Opportunities and Developments in Member States: Report on the Second Global Survey on eHealth, 2009.
[7] WHO, Telehealth, Analysis of Third Global Survey on eHealth Based on the Reported Data by Countries, 2016.
[8] C. Thuemmler, A.K.L. Jumelle, A. Paulin, A. Sadique, A. Schneider, C. Fedell, S. Covaci, 5G-PPP White Paper on eHealth Vertical Sector, EU Commission, Brussels, 2015.
[9] A. Qayyum, J. Qadir, M. Bilal, A. Al-Fuqaha, Secure and robust machine learning for healthcare: a survey, IEEE Rev. Biomed. Eng. 14 (2021) 156–180.
[10] S.M.R. Islam, D. Kwak, M.H. Kabir, M. Hossain, K. Kwak, The internet of things for health care: a comprehensive survey, IEEE Access (2015).
[11] Y. Al-Issa, M. Ottom, A. Tamrawi, eHealth cloud security challenges: a survey, Hindawi J. Healthc. Eng. (2019).
[12] I.B. Ida, A. Jemai, A. Loukil, A survey on security of IoT in the context of eHealth and clouds, in: 11th International Design & Test Symposium (IDT), 2016.
[13] C. Butpheng, K.H. Yeh, H. Xiong, Security and privacy in IoT-cloud-based e-health systems—A comprehensive review, Symmetry 12 (17) (2020) 1–35.
[14] A. Kumari, S. Tanwar, S. Tyagi, N. Kumar, Fog Computing for Healthcare 4.0 Environment: Opportunities and Challenges, Elsevier Ltd, 2018.
[15] B. Farahani, F. Firouzi, V. Chang, M. Badaroglu, N. Constant, K. Mankodiya, Towards fog-driven IoT eHealth: promises and challenges of IoT in medicine and healthcare, Future Gener. Comput. Syst. 78 (2018) 659–676.
[16] S.U. Amin, M.S. Hossain, Edge intelligence and internet of things in healthcare: a survey, IEEE Access 9 (2021) 45–59.
[17] M. Hartmann, U.S. Hashmi, A. Imran, Edge computing in smart health care systems: review, challenges and research directions, Trans. Emerg. Telecommun. Technol. (2019) 1–28.
[18] L. Sun, L. Sun, X. Jiang, H. Ren, H. Ren, Y. Guo, Edge-cloud computing and artificial intelligence in internet of medical things: architecture, technology and application, IEEE Access 8 (2020) 101079–101092.

[19] L. Greco, G. Percannella, P. Ritrovato, F. Tortorella, M. Vento, Trends in IoT based solutions for health care: moving AI to the Edge, Pattern Recogn. Lett. (2020).
[20] A. Ahad, M. Tahir, M.A. Sheikh, K.I. Ahmed, A. Mughees, A. Numani, Technologies trend towards 5G network for smart health-care using IoT: a review, Sensors (2020) 1–22.
[21] D. Zhang, J.J.P.C. Rodrigues, Y. Zhai, T. Sato, Design and implementation of 5G e-health systems : technologies, use cases, and future challenges, IEEE Commun. Mag. (2021).
[22] G. Cisotto, E. Casarin, S. Tomasin, Performance requirements of advanced healthcare services over future cellular systems, IEEE Commun. Mag. 58 (13) (2020) 76–81.
[23] H.N. Qureshi, M. Manalastas, A. Ijaz, A. Imran, Y. Liu, M.O.A. Kalaa, Communication requirements in 5G-enabled healthcare applications: review and considerations, Healthcare 10 (12) (2022) 1–33.
[24] K. Edemacu, H.K. Park, B. Jang, J.W. Kim, Privacy provision in collaborative Ehealth with attribute-based encryption: survey, challenges and future directions, *IEEE* Access (2019).
[25] R. Sanchez-Iborra, Lpwan and embedded machine learning as enablers for the next generation of wearable devices, Sensors 21 (115) (2021).
[26] O. Kocabas, T. Soyata, M.K. Aktas, Emerging security mechanisms for medical cyber physical systems, IEEE/ACM Trans. Comput. Biol. Bioinform. (2016).
[27] E. Kapassa, M. Touloupou, A. Mavrogiorgou, A. Kiourtis, D. Giannouli, K. Katsigianni, D. Kyriazis, An innovative eHealth system powered by 5G network slicing, in: Sixth International Conference on Internet of Things: Systems, Management and Security (IOT), 2019.
[28] H.B. Hassen, W. Dghais, B. Hamdi, An E-Health System for Monitoring Elderly Health Based on Internet of Things and Fog Computing, Health Information Science and Systems, Springer Nature Switzerland AG, 2019.
[29] R. Craciunescu, A. Mihovska, M. Mihaylov, S. Kyriazakos, R. Prasad, S. Halunga, in: Implementation of Fog Computing for Reliable eHealth Applications, 49th Asilomar Conference on Signals, Systems and Computers, 2015.
[30] E.L. Lydia, K. Shankar, M. Ilayaraja, K.S. Kumar, Technological solutions for health care protection and services through internet of things(IoT), Int. J. Pure Appl. Math. 118 (17) (2018) 277–283.
[31] L. Wang, Y. Hsiao, X. Xie, S. Lee, An outdoor intelligent healthcare monitoring device for the elderly, IEEE Trans. Consum. Electron. 62 (12) (2016) 128–135.
[32] J. Lloret, L. Parra, M. Taha, J. Tomas, An architecture and protocol for smart continuous eHealth monitoring using 5G, Comput. Netw. J. 129 (2017) 340–351.
[33] V. Oliveira, L. Duarte, G. Costa, M. Macedo, T. Silveira, Automation system for six-minute walk test using RFID technology, in: International Symposium on Networks, Computers and Communications, 2020.
[34] S. Ali, M. Ghazal, in: Real-Time Heart Attack Mobile Detection Service (RHAMDS): An IoT Use Case for Software Defined Networks, Canadian Conference on Electrical and Computer Engineering, 2017.
[35] G. Barbash, S. Glied, New technology and health care costs the case of robot-assisted surgery, N. Engl. J. Med. 363 (18) (2010) 701–704.
[36] Q. Zhang, J. Liu, G. Zhao, Towards 5G Enabled Tactile Robotic Telesurgery, arXiv:1803.03586, 2018.
[37] P. Sundaravadivel, E. Kougianos, S.P. Mohanty, M.K. Ganapathiraju, Everything you wanted to know about smart health care, IEEE Consum. Electron. Mag. (2018).
[38] P. Porambage, J. Okwuibe, M. Liyanage, M. Ylianttila, T. Taleb, Survey on multi-access Edge computing for internet of things realization, IEEE Commun. Surv. Tutorials 20 (14) (2018) 2961–2991.
[39] A.C. Aleixo, et al., SLICENET, Vertical Sector Requirements Analysis and Use Case Definition, 2017.

[40] Q. Wang, J. Alcaraz-Calero, M. Weiss, A. Gavras, P. Neves, R. Cale, P. Walsh, SliceNet: end-to-end cognitive network slicing and slice management framework in virtualised multi-domain, multi-tenant 5G networks, in: IEEE International Symposium on Broadband Multimedia Systems and Broadcasting, 2018.
[41] Y. Miao, Y. Jiang, L. Peng, M.S. Hossain, G. Muhammad, Telesurgery robot based on 5G tactile internet, Mob. Netw. Appl. 23 (2018) 1645–1654.
[42] D. Parry, S. Madanian, T. Norris, in: Disaster eHealth-Sustainability in the Extreme, IEEE 14th Intl Conf on Dependable, Autonomic and Secure Computing, 14th Intl Conf on Pervasive Intelligence and Computing, 2nd Intl Conf on Big Data Intelligence and Com, 2016.
[43] 3GPP, 5G; Service Requirements for the 5G System (3GPP TS 22.261 Version 16.13.0 Release 16), 2020.
[44] F. Firouzi, B. Farahani, M. Ibrahim, K. Chakrabarty, Keynote paper: from EDA to IoT e-health: promises, challenges, and solutions, IEEE Trans. Comput. Aided Des. Integr. Circuits Sys. 37 (2018) 2965–2978.
[45] F. Firouzi, K. Chakrabarty, S. Nassif, Intelligent Internet of Things: From Device to Fog and Cloud, Springer, 2020, https://doi.org/10.1007/978-3-030-30367-9.
[46] P. Afshar, A. Mohammadi, K.N. Plataniotis, Brain tumor type classification via capsule networks, in: 25th IEEE International Conference on Image Processing (ICIP), 2018, pp. 3129–3133.
[47] T. Zheng, W. Xie, L. Xu, X. He, Y. Zhang, G.Y.M. You, Y. Chen, A machine learning-based framework to identify type 2 diabetes through electronic health records, Int. J. Med. Inform. 97 (2017) 120–127.
[48] P. Varga, J. Peto, A. Franko, D. Balla, D. Haja, F. Janky, G. Soos, D. Ficzere, M. Maliosz, L. Toka, 5G support for industrial Iot applications—Challenges, solutions, and research gaps, Sensors (Switzerland) 20 (13) (2020).
[49] P. Mell, T. Grance, NIST Definition of Cloud Computing [Recommendations of the National Institute of Standards and Technology-Special Publication 800–145], NIST, Gaithersburg, MD, USA, 2011. http://csrc.nist.gov/publications/nistpubs/800-145/SP800-145.
[50] E. AbuKhousa, N. Mohamed, J. Al-Jaroodi, E-health cloud: opportunities and challenges, Future Internet 4 (13) (2012) 621–645.
[51] P. Zhang, D.C. Schmidt, J. White, G. Lenz, Blockchain Technology Use Cases in Healthcare, first ed., Elsevier Inc., 2018.
[52] Cisco, Fog Computing and the Internet of Things: Extend the Cloud to where the Things Are, 2015.
[53] M. Mahmoud, J. Rodrigues, K. Saleem, J. Al-Muhtadi, N. Kumar, V. Korotaev, Towards energy-aware fog-enabled cloud of things for healthcare, Comput. Electr. Eng. 67 (2018) 58–69.
[54] P.H. Vilela, J.J. Rodrigues, P. Solic, K. Saleem, V. Furtado, Performance evaluation of a fog-assisted IoT solution for e-health applications, Future Gener. Comput. Syst. 97 (2019) 379–386.
[55] A.M. Rahmani, T.N. Gia, B. Negash, A. Anzanpour, I. Azimi, M. Jiang, P. Liljeberg, Exploiting smart e-health gateways at the edge of healthcare internet-of-things: a fog computing approach, Future Gener. Comput. Syst. 78 (2018) 641–658.
[56] X. Fafoutis, et al., Designing wearable sensing platforms for healthcare in a residential environment, EAI Endorsed Trans. Pervasive Health Technol. 3 (112) (2017).
[57] X. Dai, I. Spasić, B. Meyer, S. Chapman, F. Andres, Machine learning on Mobile : an on-device inference app for skin cancer detection, in: Fourth International Conference on Fog and Mobile Edge Computing (FMEC), 2019, pp. 301–305.
[58] D. Mrozek, A. Koczur, B. Małysiak-Mrozek, Fall detection in older adults with mobile IoT devices and machine learning in the cloud and on the edge, Inform. Sci. 537 (2020) 132–147.

[59] M.W.L. Moreira, J.J.P.C. Rodrigues, V. Furtado, N. Kumar, V.V. Korotaev, Averaged one-dependence estimators on edge devices for smart pregnancy data analysis, Comput. Electr. Eng. 77 (2019) 435–444.
[60] D. Hou, R. Hou, J. Hou, ECG beat classification on edge device, in: Digest of Technical Papers—IEEE International Conference on Consumer Electronics, 2020.
[61] K.H. Lee, S. Member, N. Verma, A low-power processor with configurable embedded machine-learning accelerators for high-order and adaptive analysis of medical-sensor signals, IEEE J. Solid-State Circuits 48 (17) (2013) 1625–1637.
[62] S.S. Ram, B. Apduhan, N. Shiratori, A machine learning framework for Edge computing to improve prediction accuracy in Mobile health monitoring, in: Computational Science and its Applications – ICCSA, Springer International Publishing, 2019, pp. 417–431.
[63] Q. La, M. Ngo, T. Dinh, T. Quek, H. Shin, Enabling intelligence in fog computing to achieve energy and latency reduction, Digit. Commun. Netw. 5 (11) (2019) 3–9.
[64] X. Fafoutis, L. Marchegiani, A. Elsts, J. Pope, R. Piechocki, I. Craddock, Extending the battery lifetime of wearable sensors with embedded machine learning, in: EEE World Forum on Internet of Things, WF-IoT 2018 - Proceedings, 2018, pp. 269–274.
[65] M. Vuleti, V. Mujagi, N. Milojevi, D. Edge, Edge AI framework for healthcare applications, in: 4th IJCAI Workshop on AI for Ageing, Rehabilitation and Intelligent Assisted Living (ARIAL), 2021.
[66] W. Li, F. Milletarì, D. Xu, N. Rieke, J. Hancox, W. Zhu, M. Baust, Y. Cheng, S. Ourselin, M.J. Cardoso, A. Feng, Privacy-preserving federated brain tumour segmentation, in: Machine Learning in Medical Imaging, 2019. Berlin/Heidelberg, Germany.
[67] P.P. Ray, A Review on TinyML: State-of-the-Art and Prospects, Journal of King Saud University - Computer and Information Sciences, 2021.
[68] L.L.C. Google, Coral USB Accelerator Datasheet Version 1.4, G950–01456-01 / G950–06809-01, 2019.
[69] H.G. Abreha, M. Hayajneh, M.A. Serhani, Federated learning in Edge computing: a systematic survey, Sensors 22 (12) (2022).
[70] ITU, Report ITU-R M.2410–0, ITU-R M.-Minimum Requirements Related to Technical Performance for IMT2020 Radio Interface(s), ITU, 2017.
[71] M.E. Auer, T. Tsiatsos, Internet of things, infrastructures and Mobile applications, in: Proceedings of the 13th IMCL Conference, Springer, 2019, ISBN: 978-3-030-49932-7.
[72] ITU, Draft New Recommandation ITU-R M, IMT Vision - Framework and Overall Objectives of the Future Development of IMT for 2020 and beyond, 2015.
[73] M.B. Weiss, A. Gavras, P. Salva-Garcia, J.M. Alcaraz-Calero, Q. Wang, Network management—Edge and cloud computing, the SliceNet case, in: IEEE 17th Annual Consumer Communications & Networking Conference (CCNC), 2020.
[74] Y. Benslimen, J. Balcerzak, A. Pagès, F. Agraz, K. Koutsopoulos, M. Al-bado, T. Truong, Y. Benslimen, J. Balcerzak, A. Pagès, F. Agraz, S. Spadaro, Quality of Perception Prediction in 5G Slices for e-Health Services Using User-Perceived QoS, *hal-02572024*, 2020.
[75] L. Sanabria-Russo, J. Serra, D. Pubill, C. Verikoukis, CURATE: on-demand orchestration of Services for Health Emergencies Prediction and Mitigation, IEEE J. Sel. Areas Commun. 39 (12) (2021) 438–445.
[76] M.C. Soveri, Telecommunication Management; Study on Management and Orchestration of Network Slicing for Next Generation Network, 3GPP TR 28.801, Release 14, 2016.
[77] A.C. Aleixo, J. Alcaraz-Calero, M.B. Weiss, G. Bernini, G. Celozzi, R.D. Paulo, Wang, SliceNet: Enabling 5G Use Cases for Vertical Businesses, 2018.
[78] J. Baranda, J. Mangues-Bafalluy, L. Vettori, R. Martínez, K. Antevski, L. Girletti, C. Bernardos, K. Tomakh, D. Kucherenko, NFV service federation: enabling

multi-provider eHealth emergency services, in: *IEEE Conference on Computer Communications Workshops (INFOCOM WKSHPS)*, 2020.
[79] 5GT, D4.1 Definition of Service Orchestration and Federation Algorithms, Service Monitoring Algorithms, 2018.
[80] X. Wang, Y. Han, V.C.M. Leung, D. Niyato, X. Yan, X. Chen, Convergence of Edge computing and deep learning: a comprehensive survey, IEEE Commun. Surv. Tutor. 22 (12) (2020) 869–904.
[81] R. Silva, Y. Siriwardhana, T. Samarasinghe, M. Ylianttila, M. Liyanage, Local 5G Operator Architecture for Delay Critical Telehealth Applications, IEEE 3rd 5G World Forum (5GWF), 2020.

About the authors

Hela Makina She received her engineer degree in computer science from SUPCOM Higher School of communication in Tunisia, in 2007 and research MS degree (DEA) from ENIT in 2013 Junior researcher at MEDIATRON Laboratory, SUPCOM, TUNISIA.

Asma Ben Letaifa Associate Professor at SUPCOM, Higher School of communication in Tunisia Senior researcher at MEDIATRON Laboratory, SUPCOM, TUNISIA. Ph.D. in Telecommunication at SUPCOM, TUNISIA, November 2007. HDR degree in Telecommunication field from SUPCOM, TUNISIA, May 2019. Running researches on Cloud Computing, SDN and NFV, Virtualisation, Fog computing, Mobile Cloud Computing, 5/6 G, E-Health Systems, IA/ML/DL/RL/FL PUBLICATIONS: H index = 12, More than 18 journals and 50 conferences papers.

Abderazek Rachedi (S'05, M'09, SM'15) is currently working as full professor (Professeur des Universités) at the University Gustave Eiffel (UGE) and a member of the Gaspard Monge Computer Science laboratory (LIGM CNRS UMR 8049) since september 2008. He received his Habilitation to Direct Research (HDR: habilitation à Diriger des Recherches) from Paris-Est University in Dec. 2015, and his PhD degree in computer science from the University of Avignon (France) in 2008. He received his research MS degree (DEA) in computer science from the University of Savoie in France in 2003, and his engineer degree in computer science from the University of Technology and Science H. B. (USTHB) in 2002. He is author or co-author of more than 100 publications in international journals, and conferences.

CHAPTER TWO

A review on cancer data management using blockchain: Progress and challenges

Partha Pratim Ray[a] **and Poulami Majumder**[b]
[a]Department of Computer Applications, Sikkim University, Gangtok, India
[b]Maulana Abul Kalam Azad University of Technology, West Bengal, India

Contents

Advances in Computers, Volume 131
ISSN 0065-2458
https://doi.org/10.1016/bs.adcom.2023.04.002

Abstract

Cancer research has become diversified in nature. Different mechanisms and approaches are gradually being adopted and tested every now and then. Thus, a stringent or directed pathway is either not known or not paved till date. Blockchain is a recent technology that can provide trustless facility to the connected nodes acting as cancer research centers around the world to pave a directed way of consensus-centric research validation. Thus, in this chapter we first present the usefulness of blockchain for cancer care and research. Second, we provide current trends in blockchain-based cancer care. Third, we present a novel blockchain system model, i.e., BCCANA for cancer diagnosis, treatment, cure, and management. We further discuss cryptocurrencies for cancer care. Lastly, key open challenges are identified and future road map is discussed. This review presents a holistic approach to care, cure, and manage cancer by utilizing blockchain as a supportive and assistive technology.

1. Introduction

Cancer is assumed to be a collection of interrelated diseases. According to the report of the National Cancer Institute (NCI), 10 deadliest cancers including lung, colon, breast, pancreas, prostate, blood, lymph, liver, ovary, and esophagus cancer are found to be the most important causes for death in a large portion of human population each year [1,2]. It is not the case that only these 10 types of cancers occur in human body, many other organs are also get affected by the cancer. Cancer is normally characterized by uncontrolled rapid growth of cells. There may be different causes of such occurrences that range from radiation to chemical exposure to virus in the body.

Although a person has various degrees of controlling mechanisms over numerous cancer-causing agents, it has come into the notice that cancer cells and their growth are still considered as unpredictable. In some cases, they behave silently and stay mysterious for which the causes are not yet known [3]. Even after completion of a stringent treatment procedure, tricky cancer cells may hide or camouflage in any part of the body and may resurface whenever possible.

Thus, it becomes a very challenging task to encounter the cancer cell detection, diagnosis, treatment, curing, and management in real life. In majority of the cases, when it comes under the proof of existence, no practical options remain left to the medical caregivers that can be implied on the cancer patient to make his/her body cancer-free, thus resulting to unfortunate death. Conventional techniques seem to have failed to procure all the aforementioned aspects of cancer deliberation process in an efficient clinical way.

Blockchain is a recently introduced technology meant to provide decentralization, transparency, and security in a peer-to-peer communication network [4,5]. Blockchain allows users to automate consensus mechanisms to come at a common conclusion [6–8]. Healthcare services, especially cancer care, cure, and management, could be highly benefited by using blockchain-based platforms and technology stacks [9,10]. Blockchain would support cancer patients, oncologists, and insurance companies to access and process cancer-based electronic health records in seamless manner [11,12].

Preliminary studies have discussed a handful of notions to tackle specific sides of e-healthcare-related research and data management using blockchain platforms. Electronic health records-related studies are performed in Refs. [13, 14]. Investigation has been made to securely transfer the health records via blockchain [15]. Some studies show architectures to handle health records used in pharmaceutical industry [15]. In Ref. [16], a systematic review on oncology-related data sharing schemes has been discussed. Oncology-based data segregation and dissemination has been conducted in Ref. [17]. In Ref. [18], a detailed study on various use cases is performed where cancer-data-centric blockchain systems are discussed. Blockchain is also used to recycle cancer-related drug design and development [19]. A framework is proposed to demonstrate how blockchain can be useful to improve current status of radiation oncology research [20]. A nice case study is presented in Ref. [21] that discloses an efficient way of validation of cancer-based clinical trail using blockchain regulatory sandbox technique. An intensive cohort study is conducted to present improved genomic-data-based authentication on blockcha in Ref. [22]. In Ref. [23], pervasive cancer management system is proposed by using blockchain with multisensor integration.

We find that existing literature does not conform with dissemination of in-depth and detailed knowledge about blockchain-based cancer data management. Such research gap is envisaged to get catered in this literature which certainly improves knowledge base about possibility of amalgamation of cancer with blockchain. Table 1 presents the key terms and associated abbreviations.

The main objectives of this chapter can be summarized as follows:

- Discussion on the usefulness of blockchain for cancer care and research.
- Provision of current trends in blockchain-based cancer care.
- Proposed a novel blockchain-based cancer cure, research, care, and management system model, i.e., BCCANA.
- Illustration of significance of cryptocurencies for blockchain-assisted cancer research and cure.
- Blockchain-assisted cancer management technologies
- Discussion on key challenges and recommendations to the practitioners on cancer research and cure.

The rest of the chapter is organized as follows. Section 2 introduces fundamentals of blockchain and usefulness of blockchain for cancer. Section 3 addresses current trends in blockchain-based cancer care. Section 4 presents the latest inclusion of cryptocurrencies for cancer care. Section 5 outlines some use cases while using blockchain in cancer data management. Section 6 proposes architecture for cancer management using blockchain. Section 7 discusses blockchain-assisted cancer management technologies. Section 8 deals with key challenges and future road maps. Section 9 concludes the chapter.

Table 1 Abbreviation of key terms.

Abbreviation	Full form of keywords
BCCANA	Blockchain-based cancer cure, research, care, and management
AI	Artificial intelligence
BC	Blockchain
CAR	Chimeric antigen receptor
CNN	Convolution neural network
CT	Computer tomography
DApp	Decentralized application
DCL	Distributed cancer ledger
DCN	Distributed computing network
DLT	Distributed ledger technology
DNA	Deoxyribonucleic acid
EHR	Electronic health record
EM	Electromagnatism
FDA	Food and Drug Administration
FinTech	Financial technology
HIPAA	Health Insurance Portability and Accountability Act
IoT	Internet of things
KNN	k-Nearest neighbor
KSI	Keyless signature infrastructure
LBP	Lancor Blockchain Platform
NCI	National Cancer Institute
OMIS	Optomagnetic imaging spectroscopy
OTP	One-time password
P2P	Peer-to-peer
PKI	Public key infrastructure
PMX	Personal medical lockbox
PoS	Proof of stake
RAI	Radioactive iodine
RCNN	Recurrent convolution neural network
ROI	Region of interest
TCA	Trusted computing appliance

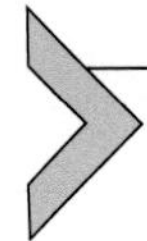

2. Fundamentals of blockchain

2.1 Basics on blockchain

Blockchain may be foreseen as a novel alternative that can be implied over the cancer data management for harnessing better results in cancer care. In generic sense, a blockchain can be inferred as a chain of continuous list of blocks that can be accessed in completely decentralized manner without intervention of any intermediator. Each block of blockchain comprises a cryptographic hash which is tied with the previous block in the same chain. A block also consists of a unique timestamp and set of transaction data which are normally organized as a Merkle tree.

The main power of blockchain lies in the facilitation of trustless service to the blockchain users where all communications are performed on a peer-to-peer (P2P) basis. Usually, the blockchain contents are immutable, i.e., unalterable; thus, data once inserted can never be deleted. At the same time, public blockchain allows the user to access the block contents at any point of time during the scope of its life time, thus maintaining a transparency [24].

Distributed ledger technology (DLT) is hereby integrated with the blockchain to ascertain that every node has same version of data [25]. Miners play the pivotal role in the working cycle of a blockchain, whereby involving consensus algorithms to validate transactions and appending a new block into the existing chain of blocks. Hence, blockchain can be summarized as an intermediary-free technology that holds high tolerance toward the famous Byzantine faults in the blockchain network. Although a blockchain can be mathematically intruded by falter nods, but in practice, it might incur computationally high expenses of processing power, capability, cost, and network knowledge that would diminish the effort of the attacker to possibility attack and take control of a whole blockchain.

Currently, four major types of blockchains exist that include public, private, consortium, and hybrid.

- Public blockchains follow "open to all" perspective to provide users access over the block contents all the time, i.e., permission-less.
- Private blockchains are normally permissioned and require invitation from the network administrator to join and access the block contents.
- Consortium blockchains use the "semi-decentralized" approach where multiple organizations can operate on a node. Thus, a restriction is implicitly imposed over the blockchain user for accessing the block data.
- Hybrid blockchains serve both purposes, i.e., public and private together.

Thus, public access over block contents is possible under the private regulations. Due to its variety and efficiency for securing data, blockchain is being utilized in different application domains, e.g., cryptocurrency trading, smart contract-based service triggering, financial technology (FinTech), supply chain, video game, space science, and healthcare [26]. However, the research community is still struggling to get the right direction and develop novel ecosystems where blockchain can be best suited.

Fig. 1 presents a generic approach to deal with cancer-based EHR using blockchain [11]. Smart contracts manage doctors and cancer patients to share secure cancer files with others. The same is made possible to be stored and accessed in a purely decentralized manner. Remote researchers, users, and pharma companies can access the cancer blocks with secure network connectivity.

2.2 Usefulness of blockchain for cancer care

Blockchain can be useful for the cancer research augmentation in coming years. We expect below-mentioned benefits that could be harnessed from the blockchains while serving cancer-related issues into reality.

- *Cancer Healthcare:* Healthcare is the most basic requirement to sustain a healthy mankind. But the situation becomes more problematic when cancer comes into the scene. Traditional clinical diagnosis and medical treatment are not coping up the cancer care successfully. Despite the fact

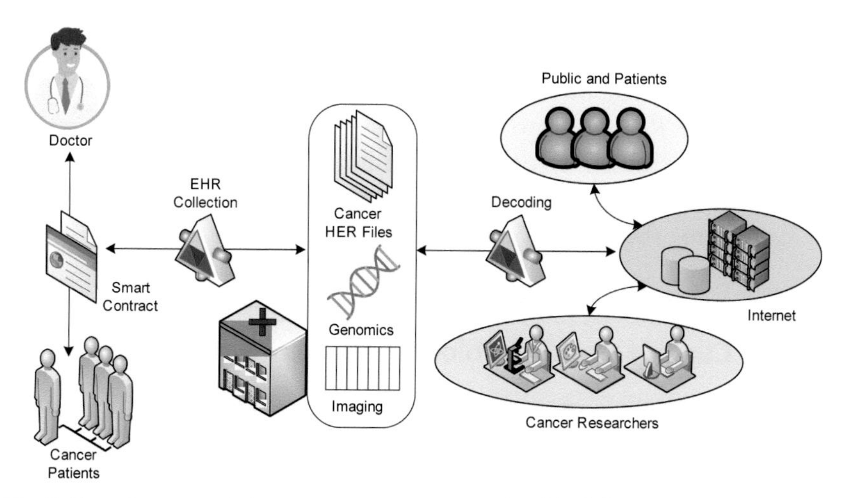

Fig. 1 Generic cancer data management system using blockchain.

that cancer is complex set of related multiple diseases where each of such diseases needs to be cured, existing clinical methods fail to support the actual need of a cancer patient.

- *Stakeholder Aspects:* Blockchain can provide a new direction toward cancer cure and research by indulging all of its specifics as mentioned earlier. A possible methodology to furnish this requirement may be realized by integrating all the three key stakeholders of current healthcare industry, such as (i) patients, (ii) medical caregiver/organization, and (iii) government agencies.
- *Cancer Ledger:* In such scenario, one can easily access the shared cancer data among the decentralized and trustless permissioned network. Same set of cancer data, i.e., distributed cancer ledger (DCL), may be readily available to each of the stakeholders that may be effectively utilized for decision-making process or a predictive purpose. Medical caregivers such as doctors, oncologists, nurses, clinic, and hospitals can access the same DCL among themselves to accurately categorize the type of cancer and the possible treatment and medications. Thus, a new option for using crowd-sourced DCL may be provisioned with help of blockchain that would support the cancer data-based research and predictive care as well as the management [27].
- *Permissioned Healthcare:* DCL under permissioned blockchain may be more efficiently utilized for streamlining of cancer dataset design and its authentication. Each of the stakeholders may significantly take part in the consensus-making process to authenticate a user/stakeholder to verify the actual identity. Hence, a transparent DCL creation process would be successfully performed whereby diminishing the possibility of fraudulent data manipulation practices.
- *Community Research:* It would help the research community to use the DCL in a trustless environment without depending upon a central authority for seeking of the authenticity of the cancer data they are using for carrying the research and prediction analytics.

3. Current trends in blockchain-based cancer care

A new set of incorporations and business models are gradually emerging into the market-place who are engaged in the proper facilitation of blockchain-based cancer care to the masses. We shall discuss the key functions of such startups [28, 29].

3.1 AH cancer blockchain

AH Cancer Blockchain has launched a consortium blockchain ecosystem that is side chain ready, i.e., cancer data are appended with pollution data, radon data, cultural data, and geographical data. The underlying artificial intelligence (AI) facilitates the correlations between the actual patient data and the appended data for in-depth screening.

Key benefits that can be levied from such ecosystem may be considered as follows (i) patient matching, i.e., professional cancer therapists can match patient symptoms, genotype, phenotype, and diagnose with other patients; (ii) treatment programs, i.e., professionals can search for the best cancer treatments being done based on these data; (iii) low cost, i.e., the blockchain connectivity is kept very cost effective; (iv) nonprofit, i.e., the program is fully performed for uplifting of cancer patients' condition, but not for profit; (v) AI-enabled; (vi) validation, i.e., all cancer treatments are validated against the privacy-aware patient data; and (vii) global usage, i.e., anybody who is consortium-approved can access the cancer data for research purpose.

3.2 ARNA genomics

ARNA Genomics is a Russian incorporation which is specialized in the development of "fluid biopsy" to early detect cancer from blood tests. It has developed the ARNA Panacea platform to monitor the efficacy of the undergoing cancer treatment by using blockchain. ARNA Panacea platform is developed for clinical trial data analysis and cancer data management to help the research community to access, store, and analyze the cancer-drug trial data over the cancer patients. The blockchain in Panacea platform is used to create a unified cancer data repository of cancer detection and medication information in crowd-source form. The platform plays a very crucial role to keep the cancer patients updated about the current observations and results obtained from the cancer research through the blockchain. It helps the professionals and researchers to store the laboratory notes, ownership details, and actual timestamp of the cancer data in the blockchain. ARNA has invented the ARNA token to pave the economy-driven ecosystem in cancer research and treatment. ARNA ERC20 tokens are helpful for streamlining cancer research through efficient utilization of the cancer clinical trials, cancer progression, cancer patient's health status, and diagnostic information over the blockchain. Anybody can use this token to buy or sell the cancer data to make the cancer research valuable for mankind.

3.3 Camelot ITLab

Camelot ITLab uses blockchain to aid in a cancer patient's treatment. The incorporation first removes the T cells from the cancer patient's body so that patients' body can produce the chimeric antigen receptor (CAR) protein. When CAR protein starts to produce, Camelot ITLab then replants the genetically modified T cells into the cancer patient's body. Thus, the goal of killing of cancer cells is achieved. It has founded the hypertrust Patient Data Care and X-Chain solution to provide the service to cancer patients. Methodologically, the CAR-T cell immunotherapy treatment is performed over the patients' body. In this case, X-Chain solution uses the blockchain technology to produce an incontrovertible audit chain for the cancer patient's data. It also verifies the cryptographical identities of each person who is involved in the process. Internet of Things (IoT)-based sensors are subsequently used to monitor and track the raw materials that are used to imply the CAR-T immunology treatment, thus ensuring that the materials are fully used by the authorized ones. The blockchain used in this solution is private; however stakeholders of this service can publicly access the chain contents. The platform also uses both blockchain and its trusted computing appliance (TCA) to make a closed-loop supply chain to store private cancer data off-chain on the TCA system.

3.4 Hybrid blockchain

Professor Dexter Hadley and his team are developing a hybrid blockchain to assist the stakeholder of health industry to detect breast cancer. They are trying to develop an optimized AI-based screening algorithm that could be trained on millions of mammograms, i.e., X-ray image of breast to allow the cancer patients all over the world to share their own mammograms over the blockchain which can only be privately accessed by such users. Their blockchain is also capable sharing mammograms, hospital data, and patients' genetic data to make the system more efficient in terms of detecting breast cancer.

3.5 Lancor Scientific

Lancor scientific has developed the Lancor Blockchain Platform (LBP) to provide a seamless journey during the cancer diagnosis and medication to the people by using the Medici crypto token. The token can be used for booking, paying, and reviewing screening tests while providing cancer patients the complete ownership on their data. Lancor has designed the opto magnetic imaging spectroscopy (OMIS) device to detect the miniscule electromagnetic (EM) changes in the cancer-affected tissue. The data are then stored into the LBP to predict the cancer up to 90% accuracy. Thus, a near real-time early cancer prediction facility can be served to anyone in the world. Users may use the Medici token to use the smart-contract features of LBP to efficiently perform transactions related to cancer screening results data, patient's personal secure profile and payment toward such services in different fiat as well as cryptocurrencies, for example, USD, GBP, Euro, Bitcoin, and Ether. Recently, Austrian government has backed the opening of a cancer research laboratory in Styria, Austria which will be run by the Lancor scientific.

3.6 Oncochain

Oncochain presents a novel approach to mitigate the blockchain study and research via creating and dissemination of a large decentralized database. Such database is de-identified about the cancer patients' data to prevent the social stigmatization via identity management. Oncochain uses a web-based oncology EHR management tool that connects cancer patients, oncologists, cancer clinics, and cancer disease-related insurance companies to work collaboratively for better cancer cure, treatment, and management. It uses innovative patient timelining mechanism to minimize the queuing delay that is required during physical checkups in the clinics. Cloud-based techniques along with blockchain and AI provide an augmented personalized and cancer-data-driven facility for effective oncological care.

3.7 OncoPower

OncoPower uses blockchain to easily share cancer data with cancer patient and cancer care team for efficient knowledge transmission about required medications, possible side effects, and actual care needed for precise patients' care. The aim is to provide transparent cancer data to both the physicians and global community who can possibly learn about the new treatments and cancer care mechanisms to (i) improve treatment outcomes, (ii) boost value, (iii) disregard waste, and (iv) make the cancer patient's experience better. OncoPower levies its power through two key components, such as (i) OncoKlinic: used by the expert oncologist, cancer physicians, and cancer service providers to login the system to harness the information about the patient's health status, responses to treatment, changes required in medication and personalized health care; predesigned smart contacts are also used by the health insurance providers to control the incentivization process to the cancer patients based on the current disease condition, and (ii) Onco-Space: it is a part of the OncoPower system where cancer patients and medical caregivers can connect with each other. Provision is also given to collaborate with field experts, family members, friends, and patient advocates for the progression of cancer treatment. All related contents and cancer information are put inside the blocks of the chain which are constantly reviewed by all stakeholders and gradually get voted and incentivized to guarantee that superlative information is offered.

3.8 Block23

It is a blockchain-based platform to facilitate "precision medication" for cancer care by knowing the way a cancer patient's genetics, lifestyle, and living environment work altogether to cause the cancer. Block23 provides an alternative to existing problems related to cancer data access, analysis, ownership, and research "moats." A decentralized registry of personal genomic and patients' health status data is utilized for analysis and prediction of cancer. Underlying tokens enable cancer patients and caregivers to store and collect the genomic health data on the blockchain. A million of population data are used to identify the statistical relationships between the collected cancer genomic data of a patient with others. Personal medical lockboxes (PMXs) leverage patients and stakeholders a private blockchain account for cancer data collection, sharing, and analysis of cancer genomic data with proper ownership and control. PMX token is used for value exchange between the cancer care giver and patients. Thus, people are incentivized as much as possible to submit and share the cancer genomic data for better cancer care for the community of cancer patients. Table 2 compares key attributes of aforementioned blockchain solutions.

4. Cryptocurrency for cancer care

We discuss three key cryptocurrencies such as Cancer Coin, CureCoin, and ONCO as the enabler of blockchain-assisted cancer management platforms. Such coins emerged from a wide range of blockchain ecosystems to mitigate service delivery issues between cancer patients and oncologists.

Table 2 Comparative analysis of blockchain platform for cancer management.

Platform	Organization	Type BC/Tools	AI	Benefits	Stakeholders
AH Blockchain	Augusta HiTech (USA)	Sidechain, consortium	Yes	Patient match, treatment, EHR validation, consortium, low-cost, not-for-profit	Government, researcher, cancer patients
ARNA Genomics	ARNA Genomic (USA)	Fluid biopsy technology	No	Breast cancer, DNA methylation solution, hybridized DNA, cancer marker	Government, patients
Camelot ITLab	Camelot (International)	Chimeric antigen receptor, X-chain	Yes	CAR-T cell, hypertrust patient data, X-chain, immunotherapy, trusted computing appliance	Patients
Hybrid Blockchain	breastwecan.org (USA)	Precision mammography	Yes	Breast cancer, mammography, secure EHR transfer, secure sharing, screening	Individual, researcher, patients
Lancor Scientific	lancorscientific.org (UK)	Lancor Blockchain Platform	Yes	Cancer screening, registry, quantum physics-based testing	All citizen
Oncochain	OncoChain Solutions (NL)	Oncology EHR	Yes	Secure patient network, web-based oncology tool, cloud, patient matching, protocol design, pharma analytics	Provider, researcher, pharma, patients
OncoPower	Witty Health Inc (USA)	Personalized chain	Yes	On-demand care, decision making, flexible scheduling, app, treatment recommendations	Patients, oncologist, pharma
Block23	Block23 (UK)	PMX Token	Yes	ShareMyinsight platform, analytics, deep learning, secure sharing	Patient, pharma, third party

4.1 Cancer coin

Cancer Coin is a newly proposed cryptocurrency by Charlie Caruso that can be used by the cancer patients and caregivers to participate in the voting process for funding of cancer research. Cancer coin is envisioned to cater the transparency of the vast pool of the distributed contributions of funds that could be maintained on the distributed ledgers. Stakeholders must vote to reach a consensus before any money could be released from the fund; thus, a transparency can be maintained.

For example, if a cancer research institute decides they would like to apply for funding from the cancer coin community, they would submit necessary funding request to the cancer coin community for possible review and approval, provided that they would constantly leverage the research progress reports. The same is applicable for a patient who wishes to get fund from a cancer charity or government toward expensive cancer treatments. Such intervention could impose greater transparency and efficiency in global drive to cure cancer. Cancer coin can significantly cut through is possible in the bureaucracy, red tape, mediators, double handling, incompetent and extravagant spending, and speediness impression and accomplishment.

4.2 CureCoin

CureCoin is a crypto token designed to reward the participants who create computing ability for the distributed computing network (DCN), i.e., Stanford University's Folding@Home. Curecoin utilizes blockchain to allow its users for free and unlimited exchange of curecoins on the decentralized network. It rewards the miners who significantly contribute their computational power to select the curecoin-based DCN by employing Proof of Stake (PoS) consensus algorithm. It also incentivizes the miners who underwrite to the strength of security of the curecoin blockchain in a environmentfriendly way. Cancer patients and caregivers can exchange the curecoin for different health services, e.g., canner-based electronic health record (EHR) storage, access, analysis, and patient's health condition monitoring in efficient manner [19].

4.3 ONCO

Onco is a cryptocurrency used by the OncoPower platform to support the growth, innovation, and strengthening of the community of cancer patients and experts of cancer care. It is exchanged when cancer services are rendered between the members of cancer community and seamlessly rewarded to the

Table 3 Comparative analysis of cryptocurrencies for cancer management.

Platform	Organization	Key attributes
Cancer Coin	Charity coin foundation	2.5-mintue block target, derived from the Litecoin, approximately 500 million total coins, total subsidy halves in 100 kilo blocks, kimo gravity well, part of CharityCoin
CureCoin	Curecoin	Folding automation, exchange, cloud mining, folding browser, distributed computing network, proof of stake, sha256 mining
ONCO	Witty Health Inc	OncoPower, OncoSpace, OncoKlinic, collaborative platform, stakeholder participation

stakeholders when service value is included to the cancer community. Adding a health service value may undergo a number of forms, such as (i) cancer specialists who are engaged in the development of a new technique of cancer treatment, (ii) cancer patients may report about the side effects of medications to health care providers, (iii) nutrition plan, (iv) counseling, and (v) oncology experts may post accurate data via voting to endorse it. Onco can further be used to provide rewards to the caregivers and exchange from health insurers [30]. As the cancer community grows by the time and more health service value is added, it is thus expected that onco could benefit every cancer patient by reducing needless expenses and advance productivity of health service providers. Table 3 compares key attributes of aforementioned cryptocurrencies.

5. Use cases in blockchain-based cancer management

In this section, we discuss about the possible use cases in the blockchain-based cancer management domain. Blockchain could be used in multimodal forms to enable cancer care, cure, and management while implying registry maintenance, cancer data access, identity management, insurance plan management, and telemedicine.

5.1 Authorized cancer data access

Cancer-based clinical and EHR data are the most important information which must be accessed by the designated authority or personnel. With the involvement of blockchain technology, cancer-related data access could be augmented with utmost diligence. Suitable private or consortium blockchain platforms might significantly improve cancer data access strategies without mishandling of sensitive information.

5.2 Aggregation of cancer registry

Cancer data are available in different forms and formats around different organizations and government agencies. Thus, effective management of such cancer data must be equipped with improved registry mechanisms. Blockchain could be used herein to aggregate various forms of cancer EHR so that overall cancer data registry process could be decentralized among its stakeholders.

5.3 Cancer patient identity management

Blockchain technology enables rudimentary approach to prevail identity management of the participating nodes in private, public, or consortium networks. Cancer patients suffer stigmatization in society. Thus, blockchain might help the cancer patients to hide their identity with secure decentralized algorithms. Moreover, transparency in blockchain network can manage the cancer patient's identity with immensely identity-aware solutions.

5.4 Cancer health insurance claim

Health insurance claim in cancer treatment is a tedious task. In such a scenario, a cancer patient (i.e., insurer) can access the prevailing rules, regulation, payment schemes, and claims in synergy with the insurance company. Thus, the chance of submitting defective or incomplete claims could be avoided with use of efficient smart contracts between the cancer patient and the insurance company.

5.5 Telemedicine care in cancer

Use of the telemedicine technology in cancer care, treatment, and management could improve the patients' overall healthcare aspects. Existing tools used in telemedicine depend on centralized cloud-based actions, making the task of regular and personalized access difficult. Blockchain might solve this use case by including blocks of payment that could be triggered when telemedicine session is over to pay the fee to the doctor. Similarly, the smart contract can be used to ping the doctor to provide necessary prescriptions to the cancer patient while removing double spending and denial of service paradigm.

5.6 Cancer data history management

Monitoring and recordkeeping of a cancer patient's medical history is another crucial job that is poorly handled in current manual scenario. Blockchain could leverage cancer-block dissemination to store medical history of a cancer patient and provide access to the designated doctor upon subsequent and successful identity management.

5.7 Cancer research streamlining

Most important use of blockchain in cancer research lies into the streamlining of cancer research. Existing research on cancer is going on in a purely random manner, thus resulting in difficulties to create a direction approach toward a better goal. Blockchain technology can provide a facility to streamline cancer research toward a specific direction while combining all the research organizations and researchers act as the trusted nodes in the consortium network. A researcher needs to pass the previous analytics which was earlier approved by consensus protocols by various trusted nodes in the network. Thus, each time a new result of research is generated, it must go through the consensus protocol so that it can be approved by the majority of the researchers. By doing so, proper streamlining of cancer research can be harnessed efficiently.

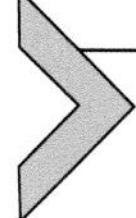

6. Proposed architecture

6.1 BCCANA

Fig. 2. illustrates the novel blockchain-based system architecture, i.e., BCCANA proposed for facilitation of cancer diagnosis, treatment, cure, management, and funding. The proposed architecture delivers cancer care services to cancer patients, individuals, pharma, cancer medical care providers, oncologists, and health insurers. The presented system integrates all the stakeholders to levy genome analysis using AI and deep learning mechanisms. The miners are allowed to get rewarded when successful consensus is reached about fund dissemination, patients' follow-up, and lifestyle management. Upon realization of the aforementioned system, cost-effective and predictive cancer care would be possible.

BCCANA architecture is proposed to mitigate the cancer care, cure, and data management under a stringent umbrella of technological intervention which is otherwise not visible in existing context of cancer care and research. BCCANA comprises six stakeholders such as (i) cancer patients,

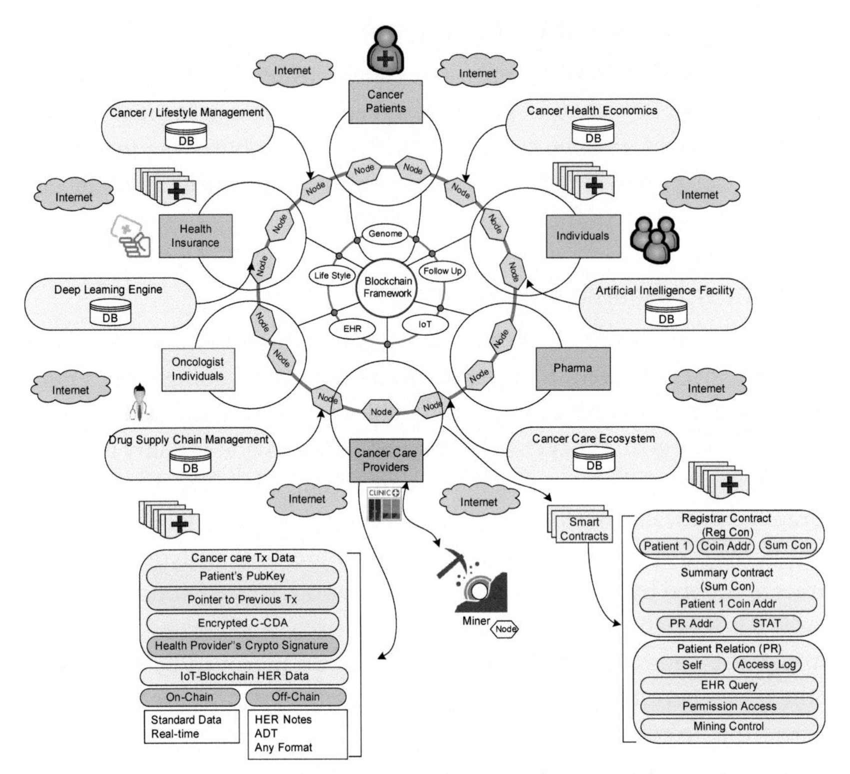

Fig. 2 BCCANA: Blockchain-based system architecture for cancer diagnosis, treatment, cure, and management.

(ii) individuals, (iii) pharmaceutical providers, (iv) cancer care providers, (v) oncologist individuals, and (vi) health insurance facilitator. All such stakeholders play a crucial role while fighting with cancer.

However, these types of stakeholders are not enough to mitigate the expected efficiency into the cancer care, cure, and data management. Thus, blockchain framework is positioned in the middle as spoke-of-wheel concept that monitors and manages everything around it. Mainly five things such as (i) internet of things (IoT), (ii) EHR, (iii) lifecycle management, (iv) genomic data provisioning, and (v) patient follow-up are strictly attached with the blockchain framework. We may better say that blockchain in the BCCANA architecture integrates all these things for more specific and direction-oriented cancer care.

All types of things and stakeholders are actually connected with the BCCANA architecture through a number of nodes. In most cases, it is connected to the blockchain framework via Internet infrastructure. Six outer layer entities are hereby juxtapose with the proposed architecture to provide secure, trustless, unbiased, democratic, and decentralized cancer-based health services; they are (i) cancer patients' lifecycle management module, (ii) cancer health economics module, (iii) AI module, (iv) cancer care ecosystem facilitation module, (v) cancer drug supply chain management module, and (vi) deep learning engine module. All these module or entities perform their designated tasks such as prediction of disease incorporation in the cancer patients' body after certain physio-bio-socio activities, cancer patients' economical condition monitoring and management, AI-based stringent cancer care provisioning, management of overall cancer care-related ecosystem, mitigating the need of patients' drug dosage, and harnessing of in-depth data learning mechanism for understanding current and future condition of cancer care, cure, and overall data management, respectively.

Indeed, miners shall be involved in every transaction related to cancer data transfer, cancer EHR facilitation, cancer patients' follow-up, and pervasive cancer care at home. In each of the cases, miners play significant roles to validate the operations in smooth and seamless manner. Normally, a set of smart contracts could be implied into the proposed framework. Communication between cancer patients, caregivers, and cancer service providers community could be paved in better way by implying peer-to-peer mechanism. IoT-based pervasive devices and cloud services could be installed at the cancer patient's end to transmit the cancer data to the trusted party without worrying about leakage of data secrecy. Fees for doctors could be given by fiat currencies; however, the miners may earn cancer cryptocurrencies as a reward.

6.2 Sequence diagram analysis

A detailed sequence diagram of the proposed BCCANA architecture is presented in Fig. 3 [10]. Fig. 3A presents the sequence of actions when a cancer patient visits the onco-clinic for the first time and meets the oncologist. If the oncologist finds that the patient gives consent to include EHR data into the blockchain platform, the oncologist then registers the cancer patient into the web-based system. An account is automatically created for the cancer patient. The cancer patient can use his/her own smartphone or personal laptop to login to the blockchain-based DApp. Impersonation is avoided by locking the account for first time entry. It creates a secure hash key which is preserved until the clinical test and diagnosis is not completed.

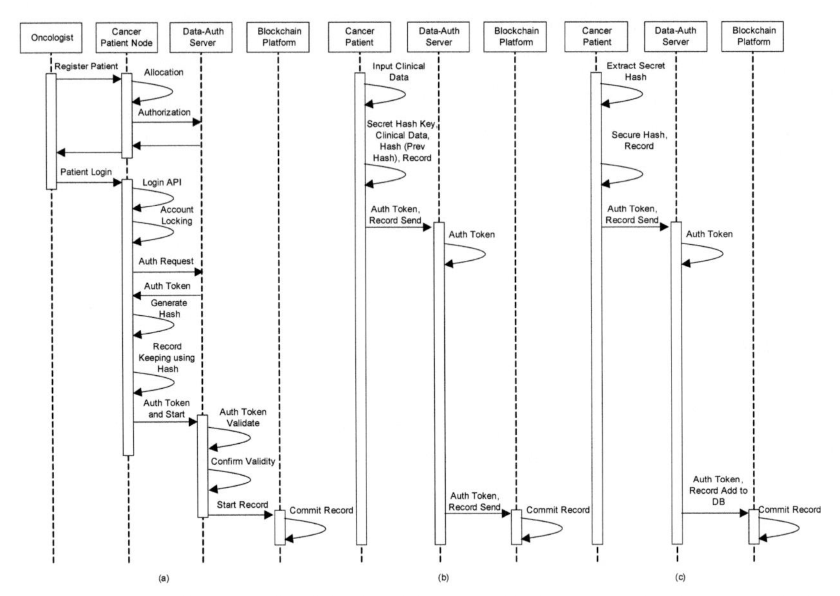

Fig. 3 Sequence diagram of BCCANA architecture. (A) Cancer patient's data inclusion, (B) during clinical diagnosis and treatment, and (C) after clinical analysis.

Cancer patient can use the DApp using his/her own credentials to store the EHR files into the tamper-proof cancer blockchain.

Fig. 3B elaborates the clinical tests and diagnosis process which is accessed by the cancer patient's personal device. The DApp first computes the hash value on the basis of the clinical test data along with the secure hash key stored in the patient's device and the previous hash. The DApp sends the patient's secure information to the blockchain with the generated hash value.

Lastly, at the end of clinical process, the DApp extracts secure hash value from the cancer patient's personal device and sends it to the secure private cancer blockchain via Internet.

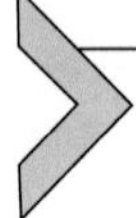

7. Blockchain-assisted cancer management technologies

7.1 Lung cancer prediction

Lung cancer can progress rapidly to result in death of the patients. Thus, early detection of lung cancer is an important factor. Use of blockchain can be helpful in early diagnosis of lung cancer. Computer tomography (CT) scan is an existing way to detect the malignancy in the lung.

But, the problem herein is associated with the delayed information about lung cancer after it has already spread. Medical attention can support the monitoring of lung cancer in a continuous manner. However, a lung CT scan image should be processed early enough by using the convolution neural network (CNN) techniques. Thus, CNN can classify and detect the lung knobs and can evaluate the possible malignancy with help of CNN-based features. The whole CNN image classification process could be integrated with blockchain technology to safeguard from any interference with the CT scan images [31]. Thus, patients can get a privacy-aware environment to establish peer-to-peer communication with the medical professional where all health-related data can be securely transmitted among them.

7.2 Breast cancer diagnosis

Breast cancer is one of the mostly susceptible types of cancer that occur among the female population. Breast cancer has become a matter of attention due to its level of occurrence among female patients. Ladies residing in rural or remote areas suffer mostly due to lesser facilities to consult with medical professionals. In underdeveloped and developing countries, we can see such issues with the female patients. Remote areas are normally suffering from infrastructure and network capabilities in these geographic areas. In Ref. [32], a deep learning-aware breast cancer diagnosis method is proposed. The proposed technique allows to minimize the lack of medical professionals and access to remote hospitals by using a 5G-assisted remote breast cancer framework. The method uses breast pathology images based on the Inception-v3 network on top of transfer learning diagnostic models. Blockchain wraps the proposed technique to make it secure, privacy aware, and mutation free.

7.3 Thyroid cancer treatment

Upon the occurrence of the thyroid cancer, the thyroid gland takes a butterfly shape near to the patient's neck. Till now, the reasons behind thyroid cancer are very less understood by the medical fraternity. However, doctors suggest that it can happen from the environmental factors as well as genetic combinations among the patients. Normally, a patient finds a lump around the neck, though not all patients face similar symptoms. The treatment of thyroid cancer often includes surgery, hormonal therapy, and radioactive iodine (RAI) induction on the patients.

Wang and Pang [33] propose an IoT-based hospital system that can help the retrospective analysis of the thyroid carcinoma (DTC) over the RAI-131 thyroid. The method can be improved with the help of IoT-enabled blockchain to provide pervasiveness, ubiquity, and privacy-supported risk factor analytics.

7.4 Cancer image sharing

In cancer management, CT scan images play an important role. Thus, image data processing should be designed in an effective manner to enable image data sharing among the patient, doctor, pathology centers, and hospital considering following factors, e.g., purely secure, mutation free, decentralized awareness, and privacy. Kumar et al. [34] propose a deep-learning-enabled secure and privacy-aware blockchain-assisted CT scan image sharing with the trusted stakeholders. The work can alleviate the privacy-aware image sharing of cancerous patients to deal with heterogeneous images. It can support the distribution of the blocked images in a decentralized secure blockchain platform throughout the world. The study aims to use the recurrent CNN (RCNN) to evaluate the region of interest (ROI) from the learned CT images stored on the blockchain.

7.5 Ovarian cancer operation staging

Women often suffer from ovarian cancer in different parts of the world. This is a very common type of cancer that is visible among the affected women. The reason lies in the inefficient methodologies to detect or diagnose ovarian cancer ahead of actual happening in the patients' body. The laparoscopy can minimize the risk of ovarian cancer by enabling short operation procedures. However, the laparoscopic surgery needs a thorough staging process before operating on the patient. In Ref. [35], an application method is designed to assist in the staging of the laparoscopy for eradication of ovarian cancer. A comprehensive staging method is proposed with the help of blockchain technology where machine learning algorithms are used to classify ovarian cancer in early stages of the operation. Blockchain enables mutation-free specifics to the laparoscopic videography. It covers the mini-camera, monitor, camera converter, cold light source, light source, path of light, and the recording components from being unauthorized accessed. Thus, blockchain helps to streamline the secure and decentralized ovarian cancer staging for a laparoscopic operation.

7.6 Cancer drug recycling

Cancer drug recycling can help patients to reuse the drugs for cancer care, cure, and management. The US Food and Drug Administration (FDA) has given permission to be a part of the cancer drug supply chain project [19]. The RemediChain [36] has started to provide recycling of cancer drugs to those patients who have monetary issues to afford them. It is trying to minimize the associated social stigma in such recycling with help of blockchain technology. Blockchain enables the transactions of cancer drugs in purely anonymous and privacy-aware manner. Also, the blockchain empowers the cancer patients to keep their identity secret and track the supply chain of the recycled cancer drugs.

7.7 Cancer genomic data sharing

Genomic data about cancer can drastically improve the drug design and cancer management tasks for the medical professionals. However, sharing of such large-sized genomic cancer data is difficult due to socio-technological barriers. Privacy preservation of a cancer patient is the most crucial aspect in this regard. We need to have a decentralized, highly scalable, and secure system to host the genomic cancer data for authorized sharing among the medical stakeholders. In Ref. [37], a blockchain system is designed that can provide electronic health records related to the genomic cancer data. This system stores the electronic records in a decentralized blockchain platform.

7.8 Occupational therapy framework for cancer patients

When a patient is diagnosed with cancer, a series of pathological tests and medications follows. Such an "after cancer treatment" requires careful monitoring to minimize the side effects from the cancer treatment aids. In Ref. [38], an IoT-blockchain-enabled occupational therapy framework is proposed to alleviate the quality of life of cancer-affected patients. The system receives data from several sensors to allow transactional records and multimedia content on top of the underlying blockchain analytic platform. The doctor can be authorized to check the patient's data in a highly secure manner.

7.9 Consensus-aware tumor diagnosis

Consensus algorithms enable the blockchain technology to take a cumulative decision to counter a given situation. Malignant tumors are susceptible to affect the patient's body organ to form a cancer. In Ref. [39], a consensus-aware blockchain platform is proposed to mitigate the risk of such tumors occurring inside the patient's breast. A consortium blockchain can be used to possess the features of the tumor in a decentralized fashion. Stringent operating standards are generated by these consensus algorithms where the proof of authority is imposed as a key tool for block creation. With the help from the cloud-based k-nearest neighbor (KNN) schemes, the tumor diagnosis is provisioned.

8. Challenges and recommendations

In this section, we discuss key challenges in existing works and how to proceed with improved strategy.

8.1 Challenges

This section provides key challenges including evolutionary aspects, cancer EHR storage, data privacy, scalability, regulatory aspects, and cancer DApp development.

8.2 Evolutionary aspects

Blockchain is a rapidly changing technology. Thus, all incorporations and allied technologies must accommodate the evolutionary aspects of blockchain. The same is applicable to cancer data management where blockchain is expected to play the most significant role in coming years. Cloud computing, IoT, edge, and data science-centric schemes need to be adjusted with the constantly changing dynamics of cancer research.

8.3 Cancer EHR storage

Cancer-based EHRs may include heterogeneous and large data formats. Thus, various ranges of data augmentation must be dealt with proper efficiency. Storage of such large amount of cancer EHR might create processing in existing infrastructure. Flexible blockchain platforms must be developed along with APIs and cryptocoins to share and access incentivization among the trusted group of peers.

8.4 Cancer data privacy

Data privacy techniques must be revisited to impose high privacy and security into the cancer EHRs decentralized in the blockchains. 5G and next-generation 6G-based communication technologies could be incorporated to strengthen cancer patients' data [40, 41]. One-time password (OTP) might be integrated with the block access and coding schemes so that privacy is completely maintained.

8.5 Scalability

Cancer blockchain solutions must cope up the scalability issue. IT should be on demand basis where decentralized nodes can participate to take decision whether it would accept new nodes or remove a node from the existing network. Such dynamic behavior should be accommodated with the blockchain infrastructure. As more number of cancer patients and clinics enroll in the network, it would increase its capacity and similarly reduce the size when a patient is cured or diseases.

8.6 Cancer DApp development

Mitigation of cancer data management requires development of decentralized application. Existing DApp is prone to volatility, efficient governance, and usability [42]. Cancer DApp must be designed in accordance with the need and user-friendly manner to handle all types of application requirements.

8.7 Design constraints

New design of cancer blockchain would face new challenges as we discussed earlier. Thus, skilled teamwork is necessary to facilitate such challenges. Industry should be accommodated with manpower with the new learning and adoption abilities to deal with cancer data management.

8.8 Regulatory aspects

Cancer data management via blockchain platforms must go through strict government laid regulations. Otherwise, there may arise malicious invasion into the secure cancer dataset. Medical regulatory agencies should develop a uniform guideline for all the stakeholders to enable themselves to harness the benefits. Social inertia should be managed simultaneously.

8.9 Recommendations

Stakeholders of cancer-related care, cure, and management activities may follow the below recommendations to better serve the patients:

- Security and privacy of cancer data is a serious matter which needs to be sincerely handled. Thus, stakeholders must use any permissioned blockchain technology in accordance to provide and get secure and privacy-aware cancer care and management services [43].
- Public blockchains are easily accessible by anyone who wants to get attached to it. Thus, it provides lesser layering over the underlying data

transmission. For cancer cure and management, private or consortium blockchain platform could be used where such issues are inherently not present.

- IoT is a proven technology that leverages smart service and pervasive application deployment in many sectors of real-life scenario. IoT could thus be considered as a great tool in the discussed ecosystem where pervasive cancer cure could be provisioned [44]. For example, a cancer patient may use IoT-based cancer data aggregation and transmitter for communicating with the peers via the blockchain platform.
- Cancer treatment incurs huge amount of money which may lead the decrease the economic freedom of cancer victims to perform essential livelihood practices [23]. Thus, knowledge of cancer economics should be incorporated within the stakeholders to control the economic condition of cancer patients.
- Stakeholders must use cancer-based cryptocurrencies to give reward to the miner who successfully mined a cancer data block with the existing system. Various such currencies are available in the market that can be easily used for leveraging efficient cancer data communication and its expenses within the ecosystem.
- Smart contracts are key to automate the blockchain-centric activities. Thus, cancer-oriented blockchain must be deployed with such smart contracts that can call some operation or other related smart contracts to fulfill the requirement of the stakeholder.
- Deep learning has recently emerged as the biggest buzz in the technology domain. Thus, deep learning techniques should be integrated with the cancer-blockchain scenario to comprehend the inner augmentation of cancer data for better service provisioning [45].
- Pharmaceutical companies manufacture different types of drugs to manage or cure cancer. However, their demand and supply are not constant. Most of the time, the price goes high, while sometimes the demanded drug as prescribed by the doctor is not available. This problem could be solved by including pharmaceutical organizations into the circuit of blockchain.
- Insurance companies must come forward and extend their support to the cancer patients and caregivers to reimburse the expenditure incurred during the tests, medicine, and management [46].
- Different research laboratories and organizations around the world should come together while forming a wheel-spoke blockchain architecture where unbiased, democratic, and stringent consensus mechanism

could be involved over the cancer EHR. Thus, the results obtained by various entities could be dynamically validated and put into the decentralized cancer ledgers across the nodes installed in the research centers [10]. In this way, a certain accuracy of result toward curing cancer may be democratically validated which will result in a better cancer solution mechanism.

9. Conclusion

In this study, we present a novel direction for augmenting cancer care, cure, and data management by using blockchain as a key enabler. Several aspects have been discussed including the importance of blockchain into the cancer care, the use of cancer cryptocurrencies, and recent trends toward blockchain-based cancer treatment. We also propose novel BCCANA architecture to integrate the stakeholders along with the goal of cancer care, cure, and research. Recommendations are appended into this chapter to motivate and guide the stakeholders of cancer-free movement to come together while practicing the given suggestions.

References

[1] P.M. Putora, et al., Oncology informatics: status quo and outlook, Oncology 98 (2020) 329–331, https://doi.org/10.1159/000507586.

[2] A.R. Arun Kumar, B.S. Aruna Kumari, C. Ramachandra, C.R. Vijay, C. Ramesh, L. Appaji, T. Avinash, N. Nelson, H.B. Govardhan, National childhood cancer comprehensive management policy-a road map, Onkologia i Radioterapia 14 (1) (2020) 6–10.

[3] R. Agrawal, S. Prabakaran, Big data in digital healthcare: lessons learnt and recommendations for general practice, Heredity 124 (2020) 525–534, https://doi.org/10.1038/s41437-020-0303-2.

[4] A.A. Mazlan, S. Mohd Daud, S. Mohd Sam, H. Abas, S.Z. Abdul Rasid, M.F. Yusof, Scalability challenges in healthcare blockchain system—a systematic review, IEEE Access 8 (2020) 23663–23673, https://doi.org/10.1109/ACCESS.2020.2969230.

[5] E. Chukwu, L. Garg, A systematic review of blockchain in healthcare: frameworks, prototypes, and implementations, IEEE Access 8 (2020) 21196–21214, https://doi.org/10.1109/ACCESS.2020.2969881.

[6] B. Houtan, A.S. Hafid, D. Makrakis, A survey on blockchain-based self-sovereign patient identity in healthcare, IEEE Access 8 (2020) 90478–90494, https://doi.org/10.1109/ACCESS.2020.2994090.

[7] A.K. Pandey, et al., Key issues in healthcare data integrity: analysis and recommendations, IEEE Access 8 (2020) 40612–40628, https://doi.org/10.1109/ACCESS.2020.2976687.

[8] R. Kumar, N. Marchang, R. Tripathi, Distributed off-chain storage of patient diagnostic reports in healthcare system using IPFS and blockchain, in: 2020 International Conference on Communication Systems & Networks (COMSNETS), Bengaluru, India, 2020, pp. 1–5, https://doi.org/10.1109/COMSNETS48256.2020.9027313.

[9] H. Im, K. Kim, J. Kim, Privacy and ledger size analysis for healthcare blockchain, in: 2020 International Conference on Information Networking (ICOIN), Barcelona, Spain, 2020, pp. 825–829, https://doi.org/10.1109/ICOIN48656.2020.9016624.
[10] A. Fernandes, V. Rocha, A.F. da Conceição, F. Horita, Scalable architecture for sharing EHR using the hyperledger blockchain, in: 2020 IEEE International Conference on Software Architecture Companion (ICSA-C), Salvador, Brazil, 2020, pp. 130–138, https://doi.org/10.1109/ICSA-C50368.2020.00032.
[11] P.P. Ray, N. Kumar, D. Dash, BLWN: blockchain-based lightweight simplified payment verification in IoT-assisted e-healthcare, IEEE Syst. J. 15 (2021) 134–145, https://doi.org/10.1109/JSYST.2020.2968614.
[12] P.P. Ray, D. Dash, K. Salah, N. Kumar, Blockchain for IoT-based healthcare: background, consensus, platforms, and use cases, IEEE Syst. J. 15 (2021) 85–94, https://doi.org/10.1109/JSYST.2020.2963840.
[13] P. Zhang, J. White, D.C. Schmidt, G. Lenz, S.T. Rosenbloom, Fhirchain: applying blockchain to securely and scalably share clinical data, Comput. Struct. Biotechnol. J. 16 (2018) 267–278.
[14] A.F. Hussein, N. ArunKumar, G. Ramirez-Gonzalez, E. Abdulhay, J.M. Tavares, V.H. de Albuquerque, A medical records managing and securing blockchain based system supported by a genetic algorithm and discrete wavelet transform, Cogn. Syst. Res. 52 (2018) 1–11.
[15] Z. Shae, J. Tsai, Transform blockchain into distributed parallel computing architecture for precision medicine, in: 2018 IEEE 38th International Conference on Distributed Computing Systems (ICDCS), IEEE, 2018, pp. 1290–1299.
[16] A. Dubovitskaya, P. Novotny, Z. Xu, F. Wang, Applications of blockchain technology for data-sharing in oncology: results from a systematic literature review, Oncology (2019), https://doi.org/10.1159/000504325.
[17] N. Dhull, J. Varghese, A.K. Dhull, Blockchain and oncology, J. Cancer Res. Ther. (2019).
[18] P. Zhang, D.C. Schmidt, J. White, Proposal for Chapter 3 Blockchain Technology Use Cases in Healthcare, 2020. Online. https://www.semanticscholar.org/paper/Proposal-for-Chapter-3-Blockchain-Technology-Use-in-Zhang-Schmidt/3f8c4cc641e92e5bc77cbfb6638528ccb839f51e.
[19] S. Benniche, Using blockchain technology to recycle cancer drugs, Lancet Oncol. 20 (6) (2019) e300, https://doi.org/10.1016/S1470-2045(19)30291-8.
[20] A. Dubovitskaya, Z. Xu, S. Ryu, M. Schumacher, F. Wang, How blockchain could empower eHealth: an application for radiation oncology, in: E. Begoli, F. Wang, G. Luo (Eds.), Data Management and Analytics for Medicine and Healthcare. DMAH 2017, Lecture Notes in Computer Science, vol. 10494, Springer, Cham, 2017.
[21] T. Hirano, et al., Data validation and verification using blockchain in a clinical trial for breast cancer: regulatory sandbox, J. Med. Internet Res. 22 (6) (2020) e18938, https://doi.org/10.2196/18938.
[22] B.S. Glicksberg, et al., Blockchain-authenticated sharing of genomic and clinical outcomes data of patients with cancer: a prospective cohort study, J. Med. Internet Res. 22 (3) (2020) e116810, https://doi.org/10.2196/16810.
[23] M.A. Rahman, M. Rashid, S. Barnes, M.S. Hossain, E. Hassanain, M. Guizani, An IoT and blockchain-based multi-sensory in-home quality of life framework for cancer patients, in: 2019 15th International Wireless Communications & Mobile Computing Conference (IWCMC), Tangier, Morocco, 2019, pp. 2116–2121, https://doi.org/10.1109/IWCMC.2019.8766496.
[24] P. Danzi, et al., Communication aspects of the integration of wireless IoT devices with distributed ledger technology, IEEE Network 34 (1) (2020) 47–53, https://doi.org/10.1109/MNET.001.1900180.

[25] C. Dookie, Building an inclusive distributed ledger system, in: 2020 IEEE 27th International Conference on Software Analysis, Evolution and Reengineering (SANER), London, ON, Canada, 2020, pp. 668–669, https://doi.org/10.1109/SANER48275.2020.9054867.
[26] D. Heaven, Bitcoin for the biological literature, Nature 566 (7742) (2019) 141–142.
[27] A. Maxmen, AI researchers embrace Bitcoin technology to share medical data, Nature 555 (7696) (2018) 293–294.
[28] L. Mertz, (Block) chain reaction: a blockchain revolution sweeps into health care, offering the possibility for a much-needed data solution, IEEE Pulse 9 (3) (2018) 4–7.
[29] C. Caruso, How blockchain could cure cancer, URL [Online]. Available: https://medium.com/@charliecaruso/how-blockchain-could-cure-cancer-e8afc0f173ef.
[30] K. Clauson, E. Breeden, C. Davidson, T. Mackey, Leveraging blockchain technology to enhance supply chain management in healthcare: an exploration of challenges and opportunities in the health supply chain, Blockchain Healthc. Today 1 (2018).
[31] S.V. Lakshmi, B. Greeshma, M.J. Thanooj, K.R. Reddy, K.R. Rakesh, Lung cancer detection and stage classification using supervised algorithms, Turkish J. Physiotherapy Rehabil. 32 (2021) 3.
[32] K. Yu, L. Tan, L. Lin, X. Cheng, Z. Yi, T. Sato, Deep-learning-empowered breast cancer auxiliary diagnosis for 5GB remote e-health, IEEE Wirel. Commun. 28 (3) (2021) 54–61.
[33] X. Wang, J. Pang, Hospital IoT system design and the efficacy of Iodine 131 in the treatment of thyroid cancer, Microprocess. Microsyst. 82 (2021) 103871.
[34] R. Kumar, W. Wang, J. Kumar, T. Yang, A. Khan, W. Ali, I. Ali, An integration of blockchain and AI for secure data sharing and detection of CT images for the hospitals, Comput. Med. Imaging Graph. 87 (2021) 101812.
[35] L. Zhang, X. Li, Y. Ning, Y. Cai, Application of laparoscopy in comprehensive staging operation of ovarian cancer based on electronic medical blockchain technology, J. Healthc. Eng. 2021 (2021) 6649640.
[36] Remedichain, 2021. Available online https://www.remedichain.org/, (accessed 31.07.21).
[37] B.S. Glicksberg, S. Burns, R. Currie, A. Griffin, Z.J. Wang, D. Haussler, T. Goldstein, E. Collisson, Blockchain-authenticated sharing of genomic and clinical outcomes data of patients with cancer: a prospective cohort study, J. Med. Internet Res. 22 (3) (2020) e16810, https://doi.org/10.2196/16810. PMID: 32196460; PMCID: PMC7125440.
[38] M. Abdur Rahman, M.M. Rashid, J. Le Kernec, B. Philippe, S.J. Barnes, F. Fioranelli, S. Yang, O. Romain, Q.H. Abbasi, G. Loukas, M. Imran, A secure occupational therapy framework for monitoring cancer patients' quality of life, Sensors (Basel) 19 (23) (2019) 5258, https://doi.org/10.3390/s19235258. PMID: 31795384; PMCID: PMC6928807.
[39] X. Zhu, J. Shi, C. Lu, Cloud health resource sharing based on consensus-oriented blockchain technology: case study on a breast tumor diagnosis service, J. Med. Internet Res. 21 (7) (2019) e13767, https://doi.org/10.2196/13767. PMID: 31339106. PMCID: PMC6683652.
[40] M. Tahir, M.H. Habaebi, M. Dabbagh, A. Mughees, A. Ahad, K.I. Ahmed, A review on application of blockchain in 5G and beyond networks: taxonomy, field-trials, challenges and opportunities, IEEE Access 8 115876–115904, https://doi.org/10.1109/ACCESS.2020.3003020.
[41] T. Nguyen, N. Tran, L. Loven, J. Partala, M. Kechadi, S. Pirttikangas, Privacy-aware blockchain innovation for 6G: challenges and opportunities, in: 2020 2nd 6G Wireless Summit (6G SUMMIT), Levi, Finland, 2020, pp. 1–5, https://doi.org/10.1109/6GSUMMIT49458.2020.9083832.

[42] R.A. Mishra, A. Kalla, N.A. Singh, M. Liyanage, Implementation and analysis of blockchain based DApp for secure sharing of students' credentials, in: 2020 IEEE 17th Annual Consumer Communications & Networking Conference (CCNC), Las Vegas, NV, USA, 2020, pp. 1–2, https://doi.org/10.1109/CCNC46108.2020.9045196.
[43] P. Zhang, D.C. Schmidt, J. White, G. Lenz, Chapter One–Blockchain technology use cases in healthcare, in: Advances in Computers, vol. 111, Elsevier, 2018, pp. 1–41.
[44] K. Lei, M. Du, J. Huang, T. Jin, Groupchain: towards a scalable public blockchain in fog computing of IoT services computing, IEEE Trans. Serv. Comput. 13 (2) (2020) 252–262, https://doi.org/10.1109/TSC.2019.2949801.
[45] F. Hughes, M.J. Morrow, Blockchain and health care, Policy Polit. Nurs. Pract. 20 (2019) 4–7, https://doi.org/10.1177/1527154419833570.
[46] O. O'Donoghue, A.A. Vazirani, D. Brindley, E. Meinert, Design choices and trade-offs in health care blockchain implementations: systematic review, J. Med. Internet Res. 21 (5) (2019) e12426.

About the authors

Partha Pratim Ray is working as active academician in the field of next generation technologies. He has published more than 100 research papers till now. He is listed as one of the top 2% scientists in the world by the Stanford University ranking. He has strong interest in conducting research in key and cutting-edge technological domain. He is presently serving as the assistant professor in the Sikkim University, India. His google scholar citations is 5300, h-index 30, i10 index 58. He is a senior member of IEEE.

Poulami Majumder has received Ph.D. in Biotechnology from Maulana Abul Kalam Azad University of Technology, Kolkata, India. She has strong interest in public health and genetics. She has published more than 30 research articles in various journals, conferences and book chapters of repute.

CHAPTER THREE

Cyber risks on IoT platforms and zero trust solutions

Marcus Tanque[a] and Harry J. Foxwell[b]
[a]Independent Researcher, Washington, DC, United States
[b]George Mason University, Fairfax, VA, United States

Contents

Advances in Computers, Volume 131
ISSN 0065-2458
https://doi.org/10.1016/bs.adcom.2023.04.003

Abstract

In recent years, organizations have developed new cybersecurity solutions, such as policies, procedures, techniques, and frameworks, to help mitigate long-term threats, risks, and potential vulnerabilities aimed at their organizational network systems. Customizing existing cybersecurity policies, processes, and standards ensures that organizations adopt and align their technology capabilities to address insider and external cyber threats. This chapter discusses Zero Trust and Internet of Things device security, protection, and interoperability. It outlines how Zero Trust concepts and methods focus on the security issues related to IoT systems. Zero Trust capability delivers security solutions for enterprise applications and endpoints. Zero Trust diverges from the traditional network perimeter, where the access request is granted without identifying and validating the requestor's message. Improved cybersecurity frameworks and tools are designed to streamline the time and resources allocated to perform IoT platforms' cybersecurity risk assessments and management frameworks. Zero Trust Network Architecture implementation focuses on robust device identity verification, compliance, and validation. This security policy is based on granted and least privileged access to only explicitly authorized resources. Continuous implementation of the Zero Trust Architecture is based on a traditional method, such as limited trust in devices connected to the organization's perimeter network infrastructure. Standardized Zero Trust policies and solutions must minimize and mitigate current and future threats, risks, and vulnerabilities. Organizations must develop and implement security best practices to ensure continuous platform federations among IoT devices/objects. Developing and implementing security best practices to ensure continuous platform federations among IoT devices/objects is key to the organization's operational functions and productivity. This chapter also discusses details about current Zero Trust solutions for IoT security. It recommends future solutions, policies, and frameworks. Zero Trust capability is innovative and advanced security architecture involving products, infrastructure, framework, strategies, vendors, and customers. Organizations are responsible for defining, enforcing, and standardizing their Zero Trust policies. This process enables monitoring and terminating the users' access and connections established between the user and enterprise resource. To adequately enforce policy, evaluating all access requests that are multi-factor authenticated is key to ensuring that proper security controls are applied. Such a process often includes all access requests submitted from an unexpected location.

1. Introduction

This chapter discusses Zero Trust (ZT) and Internet of Things (IoT) device security, protection, and interoperability. The chapter also aims to assess basic concepts, ideas, interoperability, diverse IoT cloud solutions, and in-depth reviews of current and emerging security domain solutions. Finally, it introduces a ZT architecture, strategies, and security model applied at computer security and infrastructure levels. ZT is a term that industry leaders and cybersecurity experts have used to describe a security architecture coupled with design principles and strategic methods for cybersecurity. These methods, techniques, and solutions span cybersecurity strategies, concepts, architectures, processes, procedures, and practices. In addition, ZT covers selected security strategies, frameworks, planning, and standards. These concepts define a ZT security model, frameworks, vision blueprint, and strategies for integrating device, user, network & environment, visibility & analytics, automation and orchestration, application & workloads, governance, and system backups [1,2].

In recent years, ZT transformed the enterprise IoT security paradigm. It reshaped how organizations mitigate security threats, risks, and vulnerabilities against enterprise computer resources and technology assets. In 2010, Forrester Research published ZT as a "design principle" [3–5]. Yet, despite cybersecurity and technical advances, COVID-19 Pandemic has disrupted the way leaders, cybersecurity experts, and their organizations conduct day-to-day business operations. As a result, industry leaders and cybersecurity experts are committed to evaluating and redesigning their workforce business model to address growing cybersecurity threats and data breaches and harden their corporate infrastructures [3].

Organizations rely on these design principles to establish and eliminate any implicit trust. In addition, they use these concepts as a baseline for validating digital interaction processes. The term "Zero Trust" implies that no interactions between network devices are permitted without visibility, identification, and authorization [2]. The word "Zero Trust" asserts that no devices, users, other computer resources, or technology assets on-premises or beyond perimeter security boundaries must be trusted before being verified [1]. Implementing ZT requirements such as governance, user acceptance, Privileged-Access Management (PAM), processes, and controls is complex [6]. Therefore, security practitioners must emphasize the need for software-defined perimeter benefits when deploying ZT solutions and outline frameworks and network micro-segmentation. However, organizations need more resources and expertise to ensure that ZT enterprise architecture implementation is done timely and precisely to protect physical and virtual computing resources and technology assets [1].

Implementing ZT is a continuous and complex process that requires everyone's involvement in government and industry. Such a process involves a defined and coherent roadmap for corporations and governments to move their infrastructures, applications, and data from traditional perimeter-based security landscapes to ZT perimeterless environments. For instance, stakeholders, cyber and business experts, and decision-makers must develop solutions to address ZT security gaps. Whether these security solutions are developed and implemented promptly, businesses and lawmakers must consider the risks that might inhibit productivity and how to establish responsive measures to address these concerns. In addition, the implementation of ZT must address the security gaps creation, Zero Trust Architecture (ZTA) perceptual maintenance, and insider threats, which involve risk and how to mitigate vulnerability findings. Finally, this process aims at the Zero Trust Security Model (ZTSM) to inhibit productivity which is often critical to an organization's daily operations and overall success [4,6].

The Zero Trust Security Model (ZTSM) is similarly identified as Zero Trust Network Access (ZTNA) or Zero Trust Architecture (ZTA). Zero Trust Security Model (ZTSM), Zero Trust Network Access (ZTNA), is a perimeterless security framework/borderless security that focuses on Information Technology (IT) systems design and implementation. Perimeterless security focuses on the traditional model encompassing trusted devices. These endpoints include devices that establish their connectivity via VPN. The connectivity is often initiated and established within an organization's perimeter. This concept is not ideal in today's evolving dispersed computing environments. Despite some continuing challenges, borderless security contains borders and API–Application Programming Interfaces.

In contrast, the ZT model focuses on promoting mutual authentication and verifying endpoint integrity and identity despite where they are located. Industry security experts, executives, and government policymakers describe ZTM as a "never trust, always verify." As a result, security practitioners, policymakers, and engineers argue that ZTM/ZTA devices cannot be trusted by default. Instead, all devices abide by ZTSM, and ZTA, whether connected to an authorized organization's Local Area Network (LAN), Wireless Area Network (WLAN), or other network connectivity. This security policy applies without exception, regardless of whether these devices have already been verified. These organization networks comprise interconnected zones, cloud-based service offerings, IT infrastructure, and other connections to mobile and remote network/cloud and IoT systems [7,8]. The U.S. Government establishes the ZT Trust framework, strategy, and policy to enforce trusted Internet connections, policy enforcement points, access control, adaptive access control, network access control, host containment, and verification of users and systems ZT framework comprises architecture, network, and application. In ZT architecture, verifying device identity and integrity despite the designated location to which these systems are deployed is key to a robust organization's perimeter defense posture. How these devices are mutually authenticated on the network can be determined by the level of access to applications and services delivered via the physical network, cloud, and remote endpoint interoperability. In addition, the user's confidence in devices connected via the Internet/cloud can be determined by the level of authentication and encryption capability [7,8].

In ZT, security experts define, adopt, and introduce new security measures, processes, and techniques to assess, reduce, and mitigate perimeter-based defenses, network insiders' implicit trust, and external cyber breaches. In addition, these experts continue to explore safer ways and tailored cybersecurity solutions to tackle developing perimeter-based and perimeterless security concerns. Organizational leaders and security experts must train and invest in products and solutions to enhance IoT device security. ZT spans beyond its traditional security strategies, practices, theories, governance, and techniques to protect computer resources and technology assets within perimeter-based defenses and beyond network perimeterless security. As the overarching security process implementation, organizations must ensure that their cybersecurity experts align their policies, solutions, and frameworks to ensure that detailed verification, least privilege access, and assumed breach are applied to provide 360-degree device/endpoint, user, and infrastructure protection. ZTM matured implementation stages are essential to successful solution deployment. Such process includes [7–9]:

- **Visualize**: Visualizing is based on the practical ability to be familiar with all resources needed, such as their access points and risk that can be visualized.
- **Mitigate**: Mitigating focuses on detecting and thwarting current and evolving cyber threats. It also addresses the mitigation impact of the attack or breach if an identified threat cannot be thwarted.
- **Optimize**: Optimizing is a process that encompasses the extended protection of IT infrastructure. It includes holistic protection of deployed resources despite their geographic location. Optimizing involves user experience that security and IT teams need to scale. For instance, end-users are a vital element of the holistic process when optimizing IT infrastructure protection and related resources.

Restricting every access control due to the number of access requests submitted to verify and authenticate is integral to evaluating each request before authorizing devices, users, and system connectivity per policy constraints [8,10]. ZTNA is the most favored/constructively adopted and implemented ZT model of the three: ZTNA, Zero Trust Application Access (ZTNA), and ZTA. For instance, ZT, also known as Software-Defined Perimeter, continues to be implemented as one of the robust security solutions to protect devices, networks, and data access from unauthorized users. In recent years, organizations worldwide have benefited from adopting and implementing ZT as a new global security solution. On the other hand, ZTNA is an integrated security solution that provides users access to various corporate and government infrastructures, that is, systems and accesses. The implementation of ZT focuses on a twofold approach—creating pure ZTA greenfield or hybrid ZTA solutions and perimeter systems. These methods apply to organizations that have no cybersecurity infrastructure in place. In addition, these concepts involve some organizations with existing cybersecurity defense capabilities but seeking to integrate ZTA into their perimeter-based cybersecurity systems. The chapter is divided into selected sections discussing the ZT and IoT solutions. It addresses IoT security and provides recommendations and directions for future research. Finally, we discuss how ZT impacts AWS, Azure, Google, and IBM IoT environments [3,5].

2. Background

ZT and IoT, policymakers, industry security, and management leaders continue to explore ways to improve, assess, and develop cybersecurity policies, applications, products, processes, and practices. ZT implementation requires that the government and industry collaborate to share context amid security silos and context-based policies. As an initial step toward establishing a ZT implementation strategy, decision-makers must establish a baseline to help define the architecture, policy, strategy, and security rules. The government's vision for establishing a ZT strategy is based on fundamental security concepts and requirements needed to implement such a capability. The urgent need to adopt and implement ZT must align with every organization's desire to shift from traditional to modern business. The implementation of ZT has presented numerous challenges for U.S. Government agencies and industries. These challenges include adopting and implementing the ZT framework and defining the architecture roadmap, policy, concepts, techniques, uniform standards, and strategies. To comply with ZT adoption and implementation, an organization must define the level of access to data, networks, applications, and other assets and resources.

In the post-COVID-19 pandemic era, U.S. Government agencies, partners, and corporations redefined their workforce strategy. Some have adopted a mixture of hybrid work environments. Such arrangements forced many businesses to evaluate their work arrangements to accommodate users during and post-COVID-19 pandemic. This business environment posed some challenges to agencies and industries. Currently, most employees work remotely while others continue to perform their duties at the agency and corporate sites. Establishing connectivity between employees, data, and resources through multiple IT environments planning has been one of the critical challenges for federal, state, and local agencies and enterprises to date. As a result, the government and industry have decided to adopt and implement the ZT framework as a new security solution [3–5]. ZT has compelled industries to ensure that access would be granted to users that could be trusted with accessing their infrastructure and related resources. However, despite John Kindervag's popularization of the term "Zero Trust" in 1994, Stephan Paul Marsh had a different vision of the term "Zero Trust." Despite this significant progress, the ZT architectural model still took several years. Most organizations did not begin with the slow implementation of the ZT architecture/model until a decade later. [11,12]

In 1994 Stephen Paul Marsh coined the term "Zero Trust" while working on his dissertation on computer security at Sterling University. Stephen's work focused on "trust." In the same year, Marsh introduced the "Zero Trust" concept, and his approach to "Trust" has led to what today is one of the most applied security frameworks frequently known as ZT architecture. Marsh also tailored his work on mathematical concepts, theories, and practices amid "trust." In his research, Stephen also pointed out that "trust transcends human factors." In theory, Marsh predicted that ZT would address real computer and network threats that could affect the organization's infrastructure. Such threats would consist of network systems, applications, and computing. Also, Stephen asserted that his research was based on judgment, justice, lawfulness, ethos/ethics, and morality [11,12].

In 2003, Jericho Forum (JF) was founded in the United Kingdom and introduced to the Information and Communications Technology (ICT) security standards. David Lacey, a Leading Researcher who once worked for the Royal Mail, was the visionary behind the promotion of de-perimeterization. David and his founding members' vision was aimed at discussing issues involving de-perimeterization [13]. They also submitted a Jericho Forum Commands (JFC). This set of principles highlighted how "best it was to survive in a de-perimeterized world." The JF standards were established to simplify the ICT's ability to exchange information. It also facilitates the cooperation and exchange of opinions between organizations via open-source platforms or networks [12,14].

In 2004 JF declared its first success; in 2014, it merged with The Open Group. The same year, JF identified various issues regarding drawing organizational boundaries for IT systems. While working on his commandments, JF researched, promoted, and expanded on de-perimeterization [15,16]. The original members of the JF were corporately affiliated, and the de-perimeterization solution they defined was later published as a Collaboration Oriented Architecture (COA). It focused on "securely collaborating in the clouds" [14,15]. The COA worked on the collaborative cloud approach and "identity entitlement & access management commandments" [14,17]. The intent beyond such a concept was to draft, articulate, and share its core standards and guidance. The Forum was essentially founded to help vendors produce security technology to formulate measures and guidance [14,16].

JF defines de-perimeterization as "designing the perimeters at the boundaries between an organization's ICT infrastructure/services and the open networks, individuals, and other organizations with which it connects" [14,15]. The Forum's core mission is to promote its vision among the public and private sectors. It continues to serve organizations, their contributors, and their members. It aims to develop, facilitate and inspire "security architecture and design methods" [14] involving de-perimeterization. Its central vision bridges academic and individual contributions, trading, and member-to-member collaborations [17]. The members' global responsibilities extend over three continental regions–North America, Asia Pacific, and Europe. Vendors and user members formed the JF/consortium, yet, at the outset, only user members had the opportunity to "stand for election" [14]. To some extent, the Forum discussed the trend for what today is demarcated as de-perimeterization [14,15]. In JF, de-perimeterization was defined after the members focused on the problem. The group similarly explored other variations of what has already been de-perimeterized and de-perimeterization, such as re-perimeterization, micro-perimeterization, and macro-perimeterization [14,15].

In 2009, Google introduced BeyondCorp, another ZTA version [7,8]. In 2014, "Google's BeyondCorp" promoted ZT as part of its internal architectural security concept. Despite this progress in computer security, most organizations have yet to mature their security paradigm. As a result, organizations' implementation of ZT requires a hands-on-deck strategic approach. ZT's strategic and programmatic methods rely on a thoroughly thought-based set of procedures, ideas, techniques, and concepts to help implement standardized ZTA and cybersecurity solutions. ZT is an architecture that comprises design principles [18].

In 2010, John Kindervag from Forrester Research popularized the term "Zero Trust" and its security architecture to describe and define stricter access controls and cybersecurity programs for Forrester Research [10]. John's ZT research focused on the defense-in-depth method, given its trust model that was flawed and imperfect. Kindervag pointed out, "We needed a new model that allows us to build security into the DNA of the network itself." John's research further explained that in ZT, inbound and outbound network traffic should be treated as "hostile" unless proper safety measures such as benchmarks are developed and implemented to protect security and beyond the perimeter. ZT comprises security architecture, best practices, strategies, and processes to protect Information Technology systems, infrastructures, and solutions. Industry and government implement ZT strategy to improve their IT security posture and mature current security processes, solutions, and architectures [7,8]. However, despite its contextual cornerstone, ZT's initial adoption and implementation did not happen until almost a decade after the term was coined.

In 2014 ZT network was crystallized by a Swiss Security Information Technology (IT) Engineer. The engineer's concept was merely "based on the firewall-based circuits." The engineer's idea was to design a solution to protect organizations from malware. Later, the engineer shared the architectural style manuscript with the Swiss Federal Institute of Intellectual Property (SFIIP). As a result, the same manuscript was then named "Untrust-Untrust type of network." In 2015, the manuscript was published after SFIIP's full review and completion of the manuscript. In 2019, several National Security agencies began to take notice of the ZT security evolution. Those agencies included the "National Cyber Security Centre, and UK National Technical Authority." In 2020 several platform vendors, cybersecurity, and cloud providers adopted ZT architecture into their organization's cybersecurity plan. In addition, the National Institute of Standards and Technology (NIST) and the NCSC began the ZT standardization and implementation. The essential tenets of ZT include but are not limited to–user and machine authentication, user identity, additional context, application authorization policies, and application access control policies.

These security principles have played a significant role in what today is known as the ZT security model [7,8].

In 2018 a group of cyber researchers from the NIST and the National Cybersecurity Center of Excellent (NCCoE) worked on developing the NIST Special Publication 800-207. The special publication focused on ZTA's crucial cybersecurity methods, architecture, framework/model, processes, and strategies. The publication also highlights security recommendations that U.S. Government agencies must implement to maintain and safeguard their data. It also defined ZT as a "collection of concepts and ideas designed to reduce the uncertainty in enforcing accurate, per-request access decisions in information systems and services in the face of network viewed as compromised." This enterprise architecture focuses on cybersecurity methods for utilizing ZT concepts spanning "component relations, workflow planning, and access policies." The enterprise ZT concepts generally encompass physical and virtual network infrastructure. The implementation of ZT architecture focuses on enhanced identity governance. It also on policy-based access controls, micro-segmentation, and networks, software-defined perimeters. In ZT, security mechanisms are developed to protect federal IT infrastructures, resources, assets, and enterprise network systems. Security teams incessantly monitor the organization's data usage and distribution in the ZT network [7,8].

3. Zero trust

ZT is based on the traditional models' limited security capabilities. From a holistic viewpoint, ZT embraces different security approaches and flavors. Some of these methods include creating perimeter control through complex identifies. These security and technical modalities emphasize robust authentication and identity verification of the users [19]. When most security credentials are stolen, such acts may result in security breaches that must be addressed immediately. In ZT, the need to reduce implicit trust zones and maintain availability is critical to the organization's functioning. Creating granular policies and access rules guarantees and enforces the least privileges. These privileges are needed to take actions when requested and are submitted for verification and authorization. Policy Decision Point (PDP) and Policy Enforcement Point (PEP) are integral to ZT access. How these implicit trust zones can be maintained through availability can lessen time-based delays [19].

The aim behind these security capabilities is to build a simplified, safer, and robust security architecture industry and the government needs to protect its computer resources and assets. This set of paradigms encompasses network-based perimeters, computer resources, and technology assets [5]. ZTA policies, concepts, assumptions, and techniques are built to safeguard and support computer resources and technology assets, including physical and virtual components. Sensor-based IoT devices are still vulnerable to global cyberattacks. Security experts predict that despite major cybersecurity breakthroughs in recent years, IoT devices and sensors are exposed to cyberattacks unless new solutions are developed to protect these systems at the cloud-based edge computing layer. These transformational advances have occurred due to industry investment in edge computing analytical, agility, security applications, prediction, automation solutions, and mitigation measures [3–5].

In recent months, internal and external malicious actors used the Mirai Botnet virus and other security tools to take down Internet clients: "Twitter, the Guardian, Reddit, Netflix, and CNN." Securing these network infrastructures and IoT devices requires a streamlined and long-term strategic approach that focuses on these critical requirements [18,20]:

- All devices must fall under security management policies.
- There is no distinction between "inside" and "outside" the network infrastructure.
- Access to and by IoT devices must be dynamically identifiable and observable.
- All network-attached devices must enable the highest possible defense security parameters.

These requirements aim "to prevent unauthorized access to data and services and make access control enforcement as granular as possible." The key phrase in that goal is "as possible," and therein lies the challenge. The rapidly increasing ubiquity and variety of deployed IoT devices represent one of the most challenging Internet security challenges for today's users, developers, and system architects. ZTA is a promising approach to addressing this ever-evolving challenge [20].

The National Institute of Standards and Technology (NIST) definition of this model states [3]:

> *"Zero Trust (ZT) provides a collection of concepts and ideas designed to minimize uncertainty in enforcing accurate, least privilege per-request access decisions in information systems and services in the face of a network viewed as compromised. Zero trust architecture (ZTA) is an enterprise's cybersecurity plan that utilizes zero*

trust concepts and encompasses component relationships, workflow planning, and access policies. ZTA enterprise is a network infrastructure comprising established physical, virtual, and operational policies that govern a perimeterless defense security enterprise as a product of a zero-trust architecture plan."

NIST discusses six basic ZT assumptions on network security that are associated with IoT device deployment [20]:

- The entire enterprise private network is not considered an implicit trust zone.
- Devices on the network might not be owned or configurable by the enterprise.
- No resource is inherently trusted.
- Not all enterprise resources are on enterprise-owned infrastructure.
- Remote resources and assets cannot fully trust their local network connection.
- Assets and workflows moving between enterprise and non-enterprise infrastructure must have a consistent security policy and posture.

4. Zero trust reference architecture

ZT Reference Architecture was developed alongside the "core zero trust logical component." In addition, the reference architecture includes operation and components. Such processes are developed to ensure autonomous on-premises and cloud-based services operation. ZT reference architecture focuses on a set of Enhanced Identity Governance (EIG) "crawl phase builds" capabilities, concepts, techniques, and strategies. Crawl phase builds stem from Identity, Credential, and Access Management (ICAM). This model also addresses endpoint protection. Fig. 1 outlines ZT reference architecture, processes, and workflow [21]. (See Table 1.)

4.1 NIST's ZTA core components

Over the years, score confidence has played a vital role in threat intelligence contextualization. In addition, confidence scores focus on eliminating false positives while prioritizing activities involving emerging threats. In contrast, a trust score is an objective rating that focuses on assisting organizations in mitigating their risks. Such risks often involve human interactions. This type of risk, however, may involve an individual's ethical behavior. The need for conducting enterprise policy information and access decisions amid

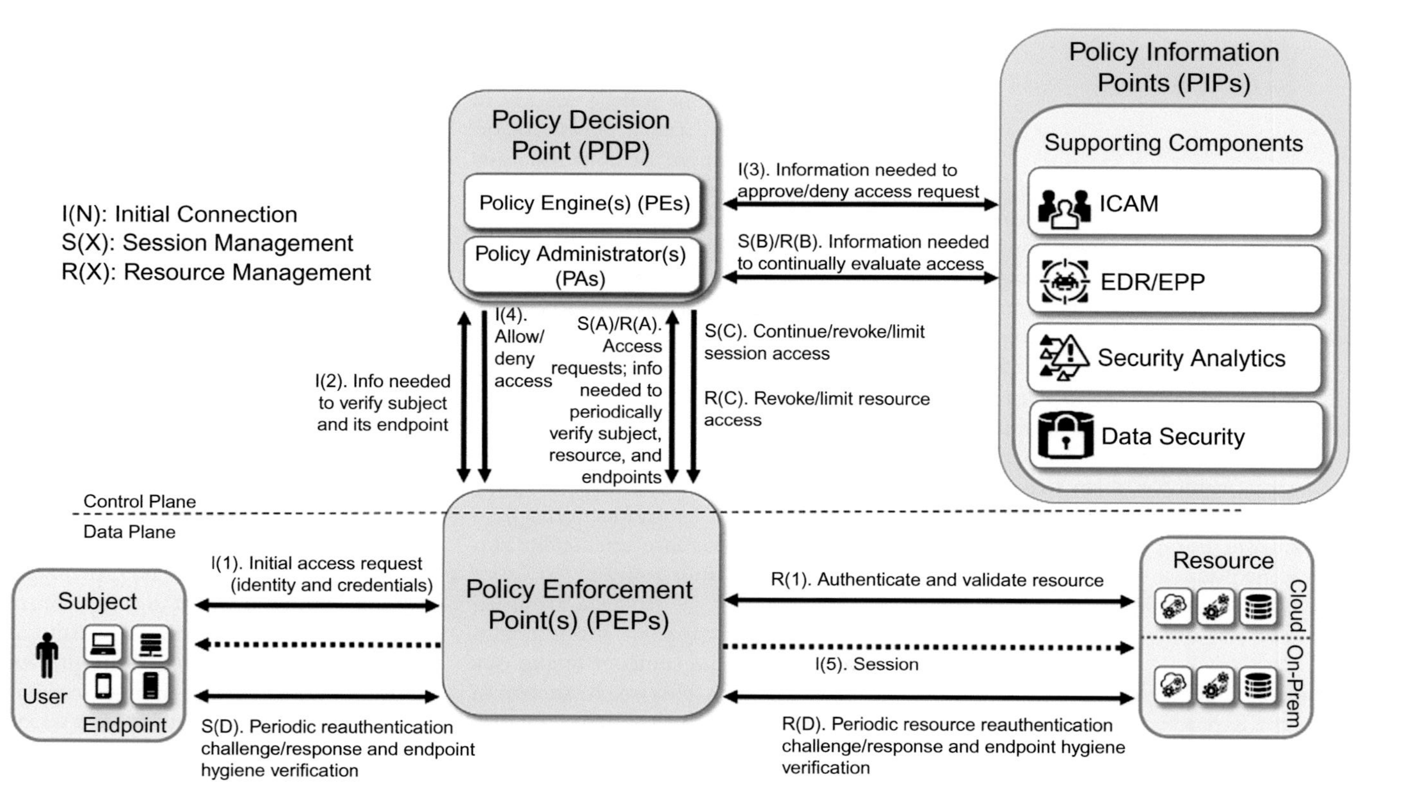

Fig. 1 NIST/NCCoE—Zero Trust Reference Architecture [21].

Table 1 Multi-factor authentication, two-factor authentication, and single sign-on [39].

Two-authentication (2FA)	Multi-factor authentication (MFA)	Single sign-on (SSO)
• It is a method that involves identity and access management. This method requires all users to present two forms of identification to access the organization's resources. It gives organizations the protection it needs to monitor and safeguard their resources, such as networks and data. • 2FA gives organizations the protection they need while preventing cyber-attacks against their infrastructure. With 2FA, users do not have to download or generate a token. • Business can use 2FA to make personalized use of its verification and identity features that are embedded into the tool. The good news about using 2FA is that passcode generators are not needed to authenticate. • It allows a maximum of password entries. This helps prevent attackers from accessing the organization's data. • 2FA is user-friendly and contains a smooth process that ensures users have minimum time to login into their systems.	• Security measure states that all users must provide single or multiple identity proofs to acquire access to a VPN application and online accounts when logging into organizations' networks or computer hardware systems. • Every organization must develop a policy that aligns with user access and verification of proof. • MFA authentication consists of three proofs of identification or other information –password, secret question, and PIN code. • Users must take steps to access their accounts, application, resources, and VPN. For example, some systems require that user access is processed through facial or voice recognition, fingerprint, thumbprint, or physical hardware essential methods. Some of these forms of authentication are required to increase an organization's security infrastructure protection against attackers and/or unauthorized users attempting to access the network.	• The single authentication allows users to gain access to their respective accounts. The access includes the organization's resources. • With SSO, organizations can grant employees access to their accounts via authentication. This method allows for user validation. In addition, organizations have the autonomous right to enable and disable their users' accounts. Once requested access is granted, users can immediately benefit from the ability to access their organization's resources through an SSO process. • SSO allows users to access their resources without re-entering new credentials. • Organizations can safeguard their properties, resources, and assets using SSO with a single solution.

Table 1 Multi-factor authentication, two-factor authentication, and single sign-on [39].—cont'd

Two-authentication (2FA)	Multi-factor authentication (MFA)	Single sign-on (SSO)
• 2FA user hardware tokens, push notifications, SMS verification, and voice-based authentication. • Hardware tokens such as key fobs randomly generate codes. Although this form of 2FA is old, most organizations in government and industry are still using it. If push notifications are used, no password is needed. Instead, a signal is generated to the user's cell phone. The user then approves or denies the massage. If accepted, then is redirected to the verification of the user's identity. • 2FA also can use text messaging to authenticate.	• The purpose of an MFA is to enhance organizational security. Therefore, users must provide valid identification before accessing their organization's online accounts, networks, or other resources. In contrast, other systems may require that users authenticate themselves through passwords and usernames. • Organizations and users must comply with the MFA policy when accessing resources. The authentication process varies form one organization to the other. Organizations require that their users authenticate to their systems using biometrics, authenticator apps such as keyfob, RSA authenticator.	• SSO allows for centralized session initiation. In addition, it ensures that user authentication services established through login credentials or security tokens are always available for users. These security features also allow users to access multiple applications with a single sign-on. Given the nature of this security capability, SSO is one of the security solutions that hackers can easily exploit internally or externally or externally to gain control of an organization's infrastructure, e.g., assets, resources, and other endpoints. • SSO provides security and compliance. It helps reduce risks involving weak, forgotten, and lost passwords. • It provides the organization with the security scalability it needs to safeguard its resources, assets, and infrastructure. • SSO can reduce any user error.

ZTA components is critical to the organization's device, network, and resource protection. There are several sectors that PE is responsible for within an organization. These sectors include single and multiple federation systems, generally known as systems of systems. In theory, every system federation is assigned a sector that must adhere to an organization's enterprise policies [9,21].

5. Zero trust architecture migration

ZT migration does not focus on replacing traditional and/or modern infrastructure. Instead, the approach is based on the organization's cultural changes. This process generally involves organization assets. Resources can be inventoried and maintained within the organization. Industries must adopt such a foundational process when adopting ZTA. Organization processes must be evaluated during the ZTA migration process. ZT principles must be implemented gradually. This implementation ensures that risk tolerance levels can be reached [22].

The adoption, implementation, or migration process involves cross-functional stakeholders. All organizations planning to adopt and/or implement ZTA must incorporate "risk-tolerance levels" into their plan [22]. Whether the implementation is greenfield or tailored toward migrating the organization's infrastructure to ZTA's current security posture should dictate the overarching approach. When implementing ZTA, "assets, resources, subjects, and processes must be identified and mapped." [22] The reason for mapping these items is to identify which candidate is ready for the ZTA implementation. The ZTA migration process includes—discovery, assessment, initial deployment, operations monitoring, and expanding the ZTA footprint. ZTA cannot be implemented as an infrastructure or process substitute. Instead, we recommend that the organization consider ZT/ZTA implementation an incremental process focusing on improving existing security infrastructure. The process must include ZT principles and develop robust processes, technology, and security solutions to safeguard data and other IT assets and resources. In recent years, U.S. Government has made significant progress in drafting policies to support the implementation and deployment of ZT solutions [22]. Fig. 2 describes the ZT deployment cycle, RMF–Risk Management Framework, and each assessment, strategy expansion, change, upgrade, and migration feedback.

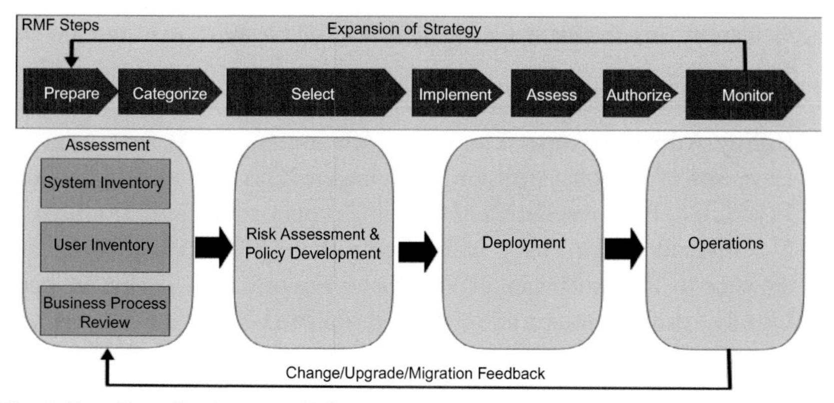

Fig. 2 Zero Trust Deployment [22].

5.1 ZT architecture core principles

ZT architecture focuses on awareness acceleration by preventing, detecting, and responding to multiple security events. Each event, however, must be handled or dealt with at least latency. If deployed along with ZT architecture, this method could be the preferred security strategy to pinpoint any scenario-based issues. We encourage organizations interested in implementing ZT architecture to adopt this NIST ZT architecture principles, framework, and processes. These Architecture Core Principles include [9,21]:

- **Policy Engine**: This Policy Engine (PE) focuses on managing last-minute issues regarding the decision to grant, revoke or deny a user's device or network access. This process includes denying, revoking, or granting resource access for a particular subject. Additionally, PE is responsible for calculating the confidence and trust scores. A confidence score is a value between zero and a hundred. This type of score allows security teams to screen through large volumes of information received from various sources while concentrating on germane threats. The need for security teams to operationalize large datasets of information regarding threats is vital to the device, network, and organization's safeguarding of its infrastructure and resources. Security teams usually look for Indicators of Compromise (IOC) that can be injected into devices and networks from multiple intelligence sources. These threats stem from a volume of "noisy data points" and related data that contains false positive outputs. To sort through such data sets and points, security teams must work painstakingly to identify any threats from ingested IOC. This effort may require that security teams have the skill to identify information needed to help decision-makers make informed decisions meticulously.

- **Policy Administrator**: The Policy Administrator (PA) focuses on the execution of the policy decision involving the ZTA's PE core component. In addition, PA is responsible for sending programming instructions, such as commands to the Policy Enforcement Point (PEP). These commands are sent to PEP to begin or end the communication session or path. Usually, this communication path is initiated amid the subject and resource. Then, it launches "session-specific authentication," credentials, and/or token authorization in the enterprise environment. The subject primarily relies on these functions to access resources within the enterprise.
- **Policy Enforcement Point**: Policy Enforcement Point (PEP) focuses on guarding the trust zone, which stores single or multiple enterprise resources. PEP manages, enables, monitors, and terminates single or multiple sessions, whether connectivity between resources and subjects in the enterprise. The full functionality of PEP relies on commands sent from the PA component. See Fig. 3 for further illustration.

The architecture in the exhibit below is complementary to the ZTA PE and PA functions comprising Policy Decision Point (PDP). PDPs issue that the subject has permission to access enterprise resources. Subjects generally comprise users and endpoints, such as devices and systems. Whereas resources generally focus on the cloud and on-premises solutions. Within the ZTA Core Components, the Policy Information Points (PIP) provide telemetry and related information required to support the PDP to "make informed access decisions." In theory, PEP is designated as the location where enforcement of access decisions takes place. The successful

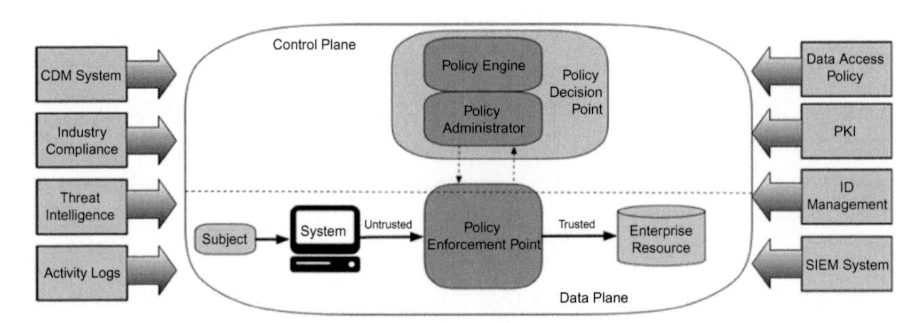

Fig. 3 NIST Zero Trust Core Logical Components [9].

functionality of ZTA is based on the Software-Defined Perimeter (SDP), EIG, and micro-segmentation. When SDP is launched gives PE the clearance to grant the subject the resource access it needs. While this activity takes place, PA performs as a network controller. It establishes a secure conduit between the subject and the resource. This channel can be established via the PEP. Fig. 3 displays the ZTA's High-level Architecture, functional components, ZTA Trust Architecture Core Components, Policy Enforcement, Policy Engine, Policy Administrator (s), and Enterprise Resources, Subject/Assets [9,21].

Security Information and Event Management (SIEM) is a vital and effective tool in securing an organization's infrastructure and resources in the ZT environment. In addition, the SIEM platform can be deployed to provide data set cross-correlation and information baseline [23]. These security processes warrant that cybersecurity teams have the tools to identify and determine data regularity and variance. These cybersecurity teams must adjust the trust level if any identified deviations exist. The appropriate response to internal or external risks and threats to the ZT environment must be addressed and mitigated through automated response action [23]. Successful implementation of SIEM in ZT strategy is significant to the deployed ZT platforms' risk mitigation and health conditions. SIEM plays a pivotal role in ZT enterprise analytics and security. A suitable implementation of the ZT approach provides clear visibility to the cybersecurity environment [23]. For instance, user behavior and activities within ZT aim to provide continuous visibility of user and system functionality, risk, and threat levels. Despite continuous interaction with ZT data, resources, and applications, organizations must consider deep visibility into the user and system as critical to the organization's interactive operational security requirements. So the dynamic functioning of ZT applications and data provisioning must be determined by the initial user and device authentication and authorization, notwithstanding more significant risk [23]. While adopting ZT, organizations must consider SIEM and security design principles. In the ZT environment, SIEM gives users the visibility they need as part of the operational cycle. In addition, a suitable ZT strategy must unequivocally comprise continuous telemetry and data log collection. In part, ZT security analytics investigation and the need to investigate the trends involved are critical to the overall ZT strategic adoption, implementation, and functional approach [23] (Fig. 4).

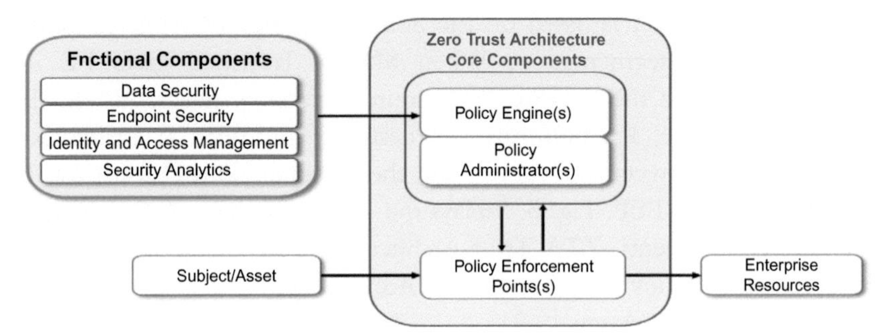

Fig. 4 NIST ZTA high-level architecture [23].

According to NIST 800-207, there are five ZTA Core Principles encompass [9,21]:

- **Multi-factor Authentication**: Multi-factor Authentication (MFA)is a security technique focusing on user access. This method aims at granting security access to both systems and networks. MFA initiates when a user(s) presents one or more pieces of evidence, such as an authentication factor to access single or multiple systems and networks. In ZT architecture, organizations and/or users use MFA to reduce identity theft incidence.
- **Least Privilege Access**: Least Privilege Access (LPA) is a method through which users are granted limited access and permissions to single or multiple networks and systems. In LPA, users are granted the least privileged permissions to access systems and networks. Users often need these permissions and access to perform their duties. ZT architecture generally relies on these security techniques to minimize the "user blast radius. The methods also ensure that critical network exposure can be minimized or dealt with accordingly.
- **Device Access Control**: Device Access Control (DAC) is a ZT architecture technique that enforces stringent device access guidelines. In DAC, device monitoring and network access are pivotal to network performance and health conditions. Therefore, all users must be verified and authorized before being granted device and network access. In addition, organizations must ensure that users have met the minimum security requirements necessary to access devices and networks before access is verified and authorized. These principles and/or techniques are required to ensure that users are verified before accessing devices and networks.
- **Micro-Segmentation**: Micro-segmentation focuses on sectionalizing security perimeters. This partitioning is based on segmenting security

perimeters into different security segments. Once segmented, these perimeters can be divided into distinct levels of workloads. Once partitioned, the perimeters must outline security controls to define service delivery for each segment. In ZT architecture, users must comply with these security controls and service-delivering processes to ensure that proper access is verified and granted. These techniques are set forth as guardrails to ensure that verified users or programs are granted strict access to a particular segment only. It also prevents users or programs from tangentially accessing other segments without separate prior authorization

- **Continuous Monitoring and Verification**: There is no interaction between users or devices in the Continuous Monitoring and Verification (CMV) technique. The internal or external user or device trust is automatically granted in CMV. ZT architecture is responsible for device verification and user identity. It also verifies privileges while ensuring periodical session time out. This process often focuses on enforcing the user and device verification cycle. Accordingly, "risk-based conditional access" warrants that the device workflow process can be interrupted should the levels of risk variation be introduced. The techniques offer continuous verification without compromising user experience.

Industry-based ZTA Core Principles include [24]:

- **Continuous Verification**: Users must always be verified before being granted access. Continuous verification is needed based on trusted zones, credentials, and devices; this process can be done anytime. When implementing ZT/ZTA, the following expression must be considered: "*Never Trust, Always Verify.*" Before adopting and implementing ZT/ZTA, the verification process must be applied.
- **Limited the "Blast Radius"**: If an insider or external attack/breach occurs, the critical impact must be minimized immediately. ZT/ZTA focuses on the restricted credential scope and access paths that attackers repeatedly use. Every attack on the systems must be mitigated and responded to timely. To limit the radius, identity-based segmentation must be applied. Implementing traditional network-based segmentation is generally a challenging operational process to maintain. For instance, data, credentials, users, and workloads are subject to constant changes. While in the least privileged principle, credentials are granted access to the minimum capacity needed to execute a task. In addition, credentials such as those of non-human accounts, e.g., service accounts, are

critical to how their access is granted. Any changes to assigned tasks must be commensurate with their scope. Often most attacks are launched through privileged service accounts. This attack process is not monitored and can exceedingly be permissioned.

- **Automate Context Collection & Response**: This principle allows for behavioral incorporation; the context of automating the collection and response must be obtained from the entire IT stack. The behavioral data collection consists of but is not limited to, the workload, endpoint, and identity. The overall process must be done accurately. Decision-makers must make accurate and effective decisions when deploying ZT/ZTA. For example, user credentials encompass non-human and human service accounts, that is, privileged accounts. This process also includes those privileged accounts that include Single Sign-On credentials. In contrast, workloads may include but are not limited to virtual machines and containers. These solutions can be deployed in a hybrid environment. However, endpoints can be classified as devices, nodes, sensors, actuators, phones, and machines accessing data in real-time.

6. Zero trust pillars

For many years risks, threats, and vulnerabilities have impacted organizations' infrastructure. Several organizations are increasingly adopting and implementing ZT architecture. The model focuses on replacing the legacy perimeter-based security. ZT security experts continue researching, developing, and deploying new security capabilities to minimize outside and insider threats, risks, and vulnerabilities disrupting the organization's day-to-day assets and resources. Vendors must develop robust and resilient defense capabilities to protect industry and government infrastructure [25]. These solutions must assist the adoption, implementation, mitigation, scalability, and maintenance. Organizations must quickly and efficiently implement and scale ZT solutions by developing such capabilities. ZTA focuses on any U.S. Government agency's IT infrastructure, that is, assets and resources protection from external and insider threats, risks, and vulnerabilities [25].

The DoD Seven ZT pillars and tents comprise [9,21]:

- **User**: This pillar focuses on the security, limitation, and enforcement of non-person and person entities requiring access to the Data, Applications, Assets, and Services (DAAS). This process is unique and requires that identity capabilities, such as "Multi-Factor Authentication (MFA) and

Privileged Access Management (PAM)," be considered in support of the privileged functions. Every organization must mature its security policy to support continuous monitoring, authentication, and authorization events. These activities must be enforced when granting user access and privileges. Decision makers must ensure that their organizations develop security benchmarks and guardrails to support MFA, PAM, and the overall security and protection between users and resources.

- **Device**: Organizations must ensure that "continuous real-time authentication" between users and devices is key to establishing security and protection across endpoints. In the ZT enterprise, device patching, assessment, and inspection must be conducted periodically with continuous real-time authentication. The introduction of a mobile device manager allows device-to-device connectivity. In theory, Trust Platform Modules (TPM), generally known as ISO/IEC 11889 standards, ought to be introduced to secure device hardware using integrated cryptographic keys. If adequately enforced, these security standards can deliver adequate device confidence assessments. Such assessments can determine the privilege for device authorization/access. Similar access requirements must be applied for each request made for the software version, encryption enablement, proper configuration, and evaluation of the compromised devices' state. As part of the ZT implementation, deployment, and resource sustainability, an organization must develop robust security requirements and solutions to protect users, resources, and assets. These solutions must determine appropriate ZT processes for identifying, remediating, securing, authorizing, authenticating, and controlling all devices in the ZT environment.
- **Network/Environment**: When deploying ZT network solutions, logical and physical segmentation continuously plays a critical role in the overall ZT network/environment. Such an approach also applies to on-premises and off-premises network environments. Whether connecting to single or multiple ZT networks, organizations must develop a granular and robust security policy to address user and device access, privilege, and network restrictions. For instance, in ZT micro/macro-segmentation, robust protection and control via DAAS are critical to the overall enterprise security. As part of the long-term ZT strategy, organizations must ensure that the following items are incorporated: privilege access, internal and external dataflow management/lateral movement prevention, and control are vital areas required to determine the level of user, device, and resource restriction and protection.

The greater the ZT perimeter becomes, the more the need for micro/micro-segmentation granularity. This method is critical when establishing robust user, device, and resource security, control, and protection measures by applying the DAAS approach.

- **Applications and Workload**: Many computer applications and workloads run on-premises. While most organizations continue to run their applications, workloads, services, and/or systems on-premises, a host of similar applications and services are stored or run on a cloud environment. For instance, ZT workloads generally comprise the entire application stack. This includes application later to the hypervisor. Therefore, organizations must secure and manage their application layer/computer containers. The security process also extends to virtual machines, which continue to be the core adoption of ZT. Proxies can well protect ZT adoption, implementation, and functioning. In ZT adoption, proxy technologies serve as application delivery that empowers ZT decisions and enforcement points. In reality, ZT adoption requires planning, agility, and flexibility. In ZT, standard/common libraries and source codes are inspected through Development Security and Operations (DevSecOps) development and practices. DevSecOps is a critical practice in ZT adoption, implementation, and functionality. DevSecOps is used in the early ZT adoption to ensure that the application development, security, and operations stages are carried out effectively from inception to securing the ZT applications.
- **Data**: The successful implementation of ZTA is based on a sheer understanding of the organization's data, applications, assets, and services, frequently known as DAAS. The need for categorizing DAAS in any organization is crucial to mission cruciality. An excellent DAAS categorization is critical to comprehensive data management development and strategy. These concepts and methods are needed to support holistic ZT adoption, implementation, and functioning. Consistently ingesting valid data and encrypting data at rest and in transit is vital in ZT adoption, implementation, and functioning. Data categorization and schemas development for data at rest and in transit are equally fundamental processes in the ZT adoption, implementation, and functioning. ZT adoption, implementation, and functioning require diverse solutioning processes. For example, ZT requires the applicability of various technology solutions, i.e., enterprise Digital Rights Management (DRM) or Information Rights Management (IRM) and Data Loss Prevention. In ZT, DRM provides encryption solutions to files and dynamically controls access privileges of files being used. In comparison, DLP focuses on

detecting any data patterns. It is also deployed to restrict any information movements that could meet specific conditions.

- **Visibility and Analytics**: As one of the core pillars of ZT, visibility, and analytics provide contextual details about the holistic performance of the environment. It is responsible for providing other ZT pillars with activity and behavior baseline. Aside from the analytics, visibility focuses on anomalous behavior detection improvement. It is also capable of making security policy dynamic changes. These changes often happen in real-time access decisions. Visibility and analytics also use different monitoring systems. In addition, it uses sensor data and telemetry to record and transmit data from remote sources to a receiving endpoint for analysis. The process can also aid in triggering alerts to produce the desired response. This process focuses on capturing and inspecting inbound and outbound traffic within the ZT environment. For instance, network telemetry and other packets play a crucial role in traffic discovery. Network telemetry can also be configured to monitor known security threats and assist in how those threats can be dealt with intelligently. In ZT, network visibility comes with enormous implications. Some of the implications include Enhanced Identity Governance (EIG). EIG is an integral method that organizations must adopt when implementing the solution. EIG focuses on the identity of the subjects, which are critical components in policy development. The EIG approach is based on access policies focusing on user device fingerprinting and identity. For example, organizations can adopt a policy where the identity of their users can be used to verify and grant access to designated resources. The policy also might dictate that micro-segmentation is essential to segmenting assets such as data centers and the cloud. This access can be granted to specific resources once the identity has been verified.
- **Automation and Orchestration**: In the Automation and Orchestration (AO) pillar, the manual security process is critical to the successful activities involved in the organization's enterprise; that is, investigation and remediation, which provides a Tier (1) Security Operations Center (SOC). In AO, speed and scalability are critical to an organization's operational posture. Using Security Orchestration, Automation, and Response (SOAR), ZT can improve its security posture while decreasing response time. For instance, security orchestration focuses on integrating SIEM. This includes the incorporation of several other automated security tools. SIEM is a vital tool crucial in handling different security systems. In ZT, automated security responses must

be handled by robust processes. Some of these processes include security policy enforcement that must be squarely established in the ZT enterprise and related environments. Fig. 5 exhibits Zero Trust Framework and each pillar process.

Subsequent DoD ZT pillars and tenets [26] were developed to complement each process exhibited in Fig. 5. See Fig. 6 for each pillar, tent process illustration, and an in-depth breakdown of each process description.

Zero Trust Pillars, Resources, and Capability Mapping focus on the Operational View (OV). The OV is intended to provide security measures that can be implemented in the ZT architecture. For instance, the Non-Person Entity identity and person identity can be monitored individually. The need for validating confidence levels through the enforcement points allows for separate paths. Authentication and authorization events often occur at enterprise-focused points. This process involves users and endpoints such as data, proxies, and applications. Analytics and SIEM receive logs from each enforcement point to develop confidence levels. Device and user confidence levels are individually collected and developed to assist in enforcing policy. Essentially, the need to authorize requested data can be determined by the non-person and person entity confidence score. The score is generally used to measure the threshold required for both entities to authorize the requested data view. Data Loss Prevention (DLP) in the ZT environment must protect data in transit. This process, however, occurs when DLP sends data to SIEM to guarantee that sent/received data is processed effectively. For a better illustration of the Pillars, Resources, and Capability Mapping, we invite you to view the below exhibit [23]. Fig. 7 illustrates the ZT pillars, resources, and capability mapping.

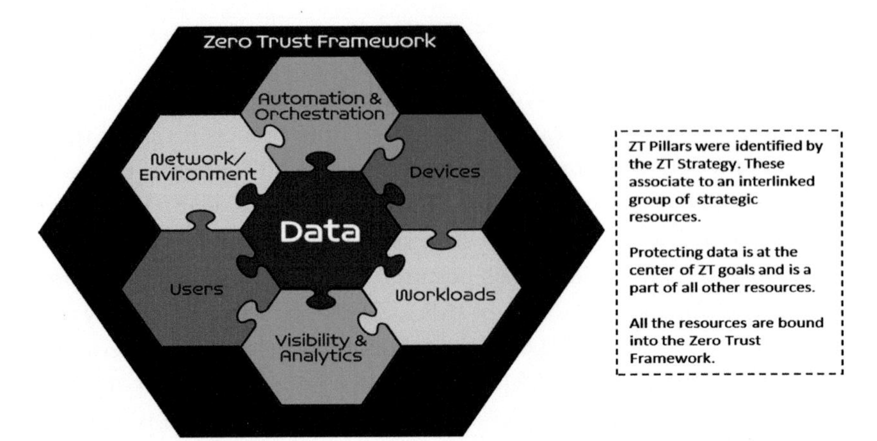

Fig. 5 DoD Zero Trust Pillars [21].

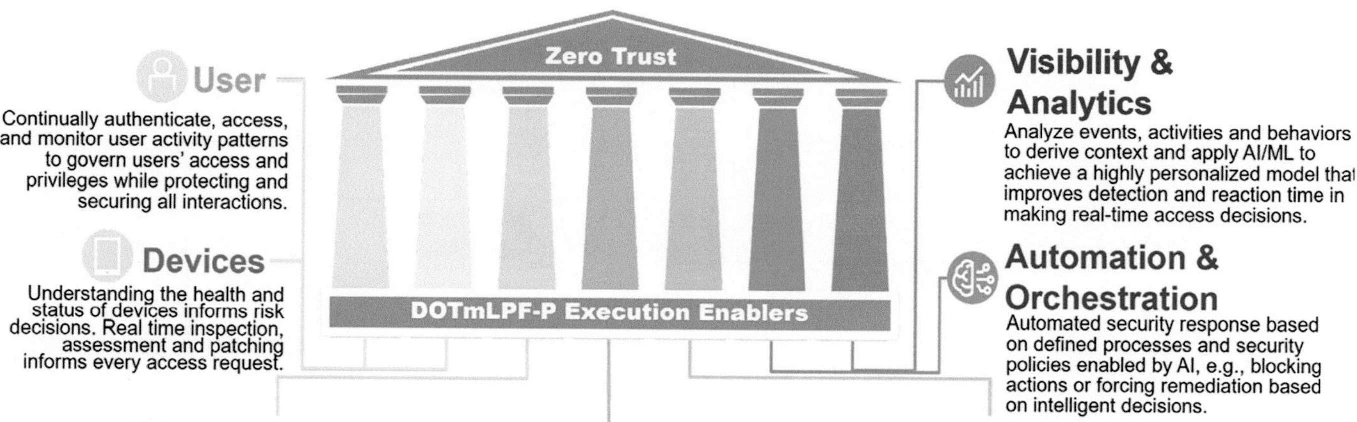
Zero Trust
DOTmLPF-P Execution Enablers
User
Continually authenticate, access, and monitor user activity patterns to govern users' access and privileges while protecting and securing all interactions.
Devices
Understanding the health and status of devices informs risk decisions. Real time inspection, assessment and patching informs every access request.
Applications & Workloads
Secure everything from Applications to hypervisors, to include the protection of containers and virtual machines.
Data
Data transparency and visibility enabled and secured by enterprise infrastructure, applications, standards, robust end-to-end encryption, and data tagging.
Network & Environment
Segment, isolate and control (physically and logically) the network environment with granular policy and access controls.
Visibility & Analytics
Analyze events, activities and behaviors to derive context and apply AI/ML to achieve a highly personalized model that improves detection and reaction time in making real-time access decisions.
Automation & Orchestration
Automated security response based on defined processes and security policies enabled by AI, e.g., blocking actions or forcing remediation based on intelligent decisions.

Fig. 6 DoD Zero Trust Pillars and Tenets [26].

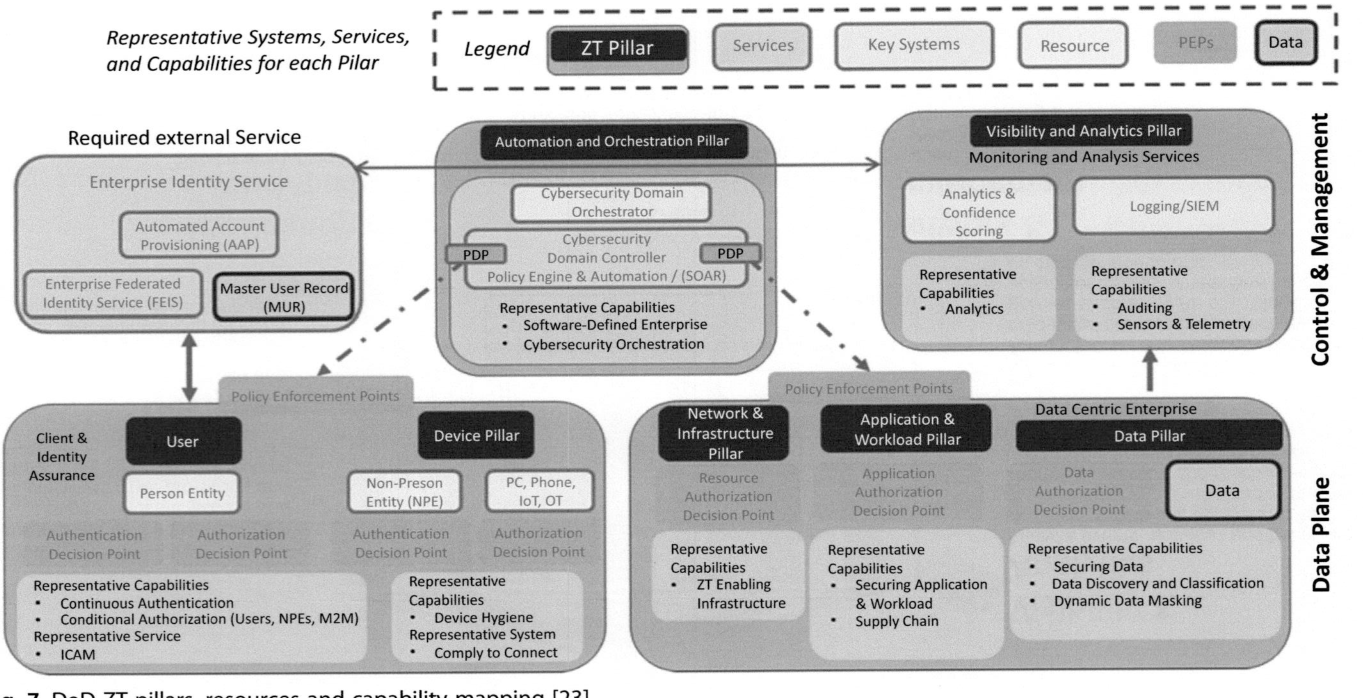

Fig. 7 DoD ZT pillars, resources and capability mapping [23].

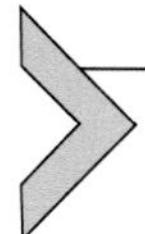

7. Zero trust network access, zero trust application access, and zero trust access

Government, corporations, and stakeholders continue to benefit from ZT technology. Yet, ZT security remains the concern of policymakers, users, and organizations. As a result, decision-makers and cybersecurity experts must develop security processes to address continued threats, risks, and vulnerabilities against the organization's devices and infrastructure. In ZT, access must be managed via secure remote access to assigned devices and networks. This process involves transitive trust that must be established to validate and/or deny device or network access in case of any threat, risk, or security breach.

The ZT framework encompasses three distinctive models—Zero Trust Network Access, Zero Trust Architecture, and Zero Trust Application Access. These models are designed to protect the network, application, data, and users. Organizations must control their device, user, and network protection level. Such a level of protection includes infrastructure security posture. Decision-makers and cybersecurity experts must determine which security model to apply to align with their security posture requirements.

In ZTNA, all access to organizations' assets, resources, and systems must be verified and authenticated. As one of the preferred ZT security solutions, ZTNA implementation focuses on network micro–segmentation and isolation. The authors predict that as a security capability, ZTNA is intended to replace Virtual Private Network (VPN) solutions in years to come. It aims to provide users access to data, networks, and applications remotely or from different geographic areas. This process also gives users the ability to access data, applications, and networks from various locations providing that they are verified and authenticated in advance. Before implementing the ZTNA model, stakeholders must carefully evaluate ZTNA, ZTAA, and ZTA's advantages and disadvantages. In an advanced ZTNA environment, cloud brokers are responsible for initiating or sending messages to users and devices. This includes those devices and users dispersedly located in various internal or external geographic networks. In theory, users and devices must be authenticated before access is granted [13].

ZTNA Advantages:

- Blocking intruders' ability to access devices and infuse malware within the ZT perimeter fully.
- Replacing legacy VPN solutions mainly in today's age, where most of the global workforce either works remotely or hybrid.

- Protecting and providing external networks with secure access is needed to protect the infrastructure, devices, and related assets.

ZTNA Disadvantages:

- Providing Apps with no protection against hackers.
- Users must trust their provider(s).
- It does not provide secure access to data, applications, and networks while users are offline. This approach includes establishing access from Operational Technology, Systems Applications and Products, and Enterprise Resource Planning.

ZTAA—Focuses on similar principles as ZTNA. However, ZTNA is a principle based on the assumption that networks are usually compromised. Should the ZTNA be compromised, ZTAA assumes it should grant access to its applications as soon as all devices and users have been authenticated.

ZTAA Advantages:

- It blocks cyber-attackers from accessing the network.
- It protects apps within the enterprise and those installed on users' devices.

ZTAA Disadvantages:

- For optimal functioning, ZTAA must establish end-to-end trust with its provider.
- If devices are offline, ZTAA establishes secure access to close network, i.e., OT, SAP, and ERP.

ZTA—This technique ensures that the ZT establishes end-to-end access transversely to all networks. This access includes apps, systems data centers, and related components within the enterprise. ZTA is exclusively based on identity access. The true ZT is not constrained by any specific platform(s) or network(s). As such, ZTA focuses on holistic ZT across an organizational architecture. In ZTA, offline networks such as ERP and the industrial floor often operate as a complete ZT capability. The holistic capability of ZTA range from ZTNA and ZTA models. ZT is an identity-based model, and its approach can also be applied to the ZTA provider. In ZTA, users do not have to establish trust before accessing the environment. In addition, ZTNA and ZTAA providers can directly access the organization's information and validate access. If an organization is concerned with data privacy, ZTA is the preferred model. In ZTA, inbound network traffic is not permitted. Therefore, secrets and customer data can be stored in the ZTA component deployed to the organization's secure environment. If kept there, the ZT provider cannot access such information. Stored data includes but is not limited to policies and password vaults. Customer keys such as certificates can be stored in the same secure environment, and unauthorized users cannot access such data [13].

ZTA Advantages:

- It prevents cyber-attackers, such as hackers, from accessing the network architecture.
- It conceals all networks and protocols.
- ZT providers cannot access customer data. This means that any data stored in the network cannot be accessed.

ZTA Disadvantages:

- Organizations must change their approach regarding networks and VPNs.

Organizations must adopt innovative transformations amid ZT implementation. This approach involves network architecture requiring a level of transformation spanning security methods and others. Studies indicate that ZT can replace the current VPN solution if fully implemented and deployed. In so doing, organizations must change their culture by adapting to ZT innovation and implementation trends. ZT is the future and allows users to establish connectivity to multiple devices and networks in the enterprise. Decision-makers must decide which security model aligns with their long-term strategic, tactical, and operational vision. The need for managers and executives to decide which of these security models, i.e., ZTAA, ZTNA, and ZTA, to adopt, deploy and implement are critical to the organization's long-term success.

8. Zero trust for IoT security

Zero Trust is established and embedded in the "never trust, always verify." The need to deliver the least privilege at scale approach by setting role-based access controls is key to successful ZT implementation. This approach includes but is not limited to continuous monitoring, analyzing, and adapting optimal strategies to address existing and emerging threats, risks, and vulnerabilities. Additionally, organizations must have 360-degree visibility of ZT environments. Such a visibility includes attributes, permissions, entitlements, and roles. When discussing trust and protection of perimeter-focused network defense and beyond settings, leaders and security practitioners such as the Chief Information Security Officers must consider how to digitally transform and take advantage of ZT security capabilities and strategies such as network segmentation to develop a robust plan that aligns with the organization's modern cybersecurity playbook [4,27].

Organizations must include Identity and Access Management when adopting and implementing ZT security. These concepts provide other capabilities, such as assessing and updating data in transit. For instance, these methods must be applied through identity lifecycle automation with real-time response to access risks. The process includes access automation, identity management, integrity, most minor privilege access/management, and continuous monitoring while responding to current or emerging cyber threats, risks, and vulnerabilities that may target the infrastructure. Organizations must invest in more advanced Artificial Intelligence (AI) and Machine Learning (ML) applications for optimal results. AI and ML solutions can provide organizations with the identity intelligence they need to predict and mitigate insider and external threats. Automating user access and updating roles when employees separate from the organization is key to the process [27]. Robust security, protection, and interoperability in the enterprise network and data center environment are vital to computer security and device management [27].

IoT is a term that focuses on wide-ranging, dispersedly connected, and distributed machine-to-machine, for instance, intelligent devices and objects. Such technologies include wearables, controls/m, sensors, and appliances. These devices are built to collect, share, analyze, and transport data from one device to another endpoint. IoT devices ensure that data is accessible, transmitted, disseminated, and extracted. Yet, despite these technological advances, IoT systems pose significant threats, risks, and vulnerabilities to society. These systems are prone to cyber-attacks and data breaches [4,27].

ZT requires a hands-on-deck approach that involves interdepartmental leaders, teams, and security practitioners to implement a unified strategy. Such implementation, however, may need a rigorous process that incorporates IoT devices and non-IoT requirements. Organizations must adopt a ZT security model that meets their physical and virtual device requirements. This model must address how organizations protect their users, applications, data, and systems. Established securities policies, standards, and procedures ensure that organizations have the resources to defend their perimeter-focused network defense and beyond. By adjusting to ZT frameworks, concepts, techniques, and other existing cybersecurity capabilities, organizations can close any security gaps and manage threats, risks, and vulnerabilities. Organizations must establish policies addressing the "never trust, always verify" methods at all times [18]. When implementing ZT, organizations must carefully review all requests for access to their physical and virtual resources before granting authorization. Security vendors have developed encryption software to protect the infrastructure from insider and external threats [27].

Security experts must create rules that address micro-segmentation and least privileged access. Organizations need these security principles to minimize any lateral movements in the enterprise. Although IoT is a widely referenced technology in the industry, its term is often misunderstood for a collective of network components such as environmental sensors and small devices embedded in homes, automobiles, residential, and industrial facilities. The word spans urban control systems, such as mobile or fixed location devices with an Internet Protocol, also known as IP address. IoT devices collect data and report it to a central service over the public Internet. They include "things" such as devices or objects found in residential and office buildings, machines, traffic control sensors, embedded & wearable personal health monitors, and other intelligent appliances. Because these devices generally use microprocessors with limited power and capabilities, they cannot include management applications such as security software [20,28]. These computer resources and technology assets can lead to severe vulnerabilities and large-scale network attacks like malware exploitations such as the Mirai Botnet [29]. Segmenting the network within a perimeter and beyond is necessary to deter lateral movement. Network segmentation provides layer seven, which comprises threat prevention simplified granularity and least access strategies [2,18,20].

Emerging commercial and open-source IoT systems pose perimeterless defense security concerns. AWS, Azure, and Google are among the leading technology companies that widely ZT-deployed IoT devices, services, and the cloud in solutions on-premises, residential, and offices. Cybersecurity applications are integral to the security and protection of an organization's computer resources and technology assets. These applications are built to help mitigate different levels of security, threats, risks, and vulnerabilities to the perimeter-focused network defense and perimeterless cybersecurity postures. Full-fledged deployment and configuration of these applications give users, devices, and beyond the enterprise extra security capability for protection. ZT incorporates design principles and security requirements to protect perimeter-focused network defense and perimeterless cybersecurity systems in physical and virtual environments. ZT secure environment spans–network, infrastructure, data, applications, and devices [2,20].

9. Zero trust endpoint security for IoT platforms

Organizations must protect their networks, devices, and users from an insider, external, and adversarial cyber threats, attacks, risks, and vulnerabilities. ZT endpoint, IoT device/network access, and credential assignments must be established to protect the entire infrastructure. In so doing, organizations must establish policies that adhere to IoT devices and endpoint access. Implementing ZT capability provides robust security controls and ensures that these devices are equipped with embedded solutions to discover, track, mitigate, enforce, and verify every user's access to the network. Such defensive mechanisms help leaders make informed decisions for protecting IoT devices and endpoint environments. Endpoint security for ZT solutions and IoT platforms has increased in scale and requirements. As a result, businesses are developing new IT security strategies to mitigate current and future threats, risks, and vulnerabilities. Despite significant progress that has been made, security vendors admit that malware viruses are the biggest threats to endpoints. The lack of robust solutions to ensure that user credentials and network access are protected remains one of the main concerns when establishing security strategies for the endpoint. ZT model goes beyond endpoints and IoT device implementation. The model has been deployed on other enterprise projects, such as network communications, to protect the infrastructure from known and unknown adversarial attacks [30].

Advanced Endpoint Protection (AEP) solutions and cyber capabilities, e.g., ZT, have been deployed to thwart malicious attacks. The unified ZT and Endpoint capability aims to provide industries with an extra security layer to protect network applications and data. Yet, endpoints are vulnerable to insider, external and adversarial cyber-attacks. AEP blocks unidentified malicious attacks that are injected into the endpoints. The malicious attacks can be launched as malware that hackers send to endpoints to trigger zero-day attacks. Through zero-day exploits, hackers can launch cyber-attacks on various chosen targets. These types of cyberattacks allow hackers to steal data and exploit endpoint vulnerabilities. Once the vulnerabilities are identified must be mitigated within zero days. These days give security practitioners the ability to scan and fix system vulnerabilities. So, security vendors must build solutions to detect and mitigate threats against endpoints and IoT devices. Zero-day is a new "uncovered software vulnerability" [30]. Once software developers have identified these vulnerabilities, they have zero days, aka 0 days, to scan and find remediation to the issue. However, there is always a possibility that hackers might regain or exploit the endpoints after being patched [30].

There are three types of zero-day methods [30]:

- **Zero-day Threat/Vulnerability**: This type of threat/vulnerability is best described as an unknown security flaw/defect that software developers can find when scanning the endpoint.
- **Zero-day Exploit**: A method that attackers use to exploit and gain control of zero-day weaknesses. These types of vulnerabilities can be exploited for mischievous intents, e.g., malware that can be injected into endpoints to trigger cyber-attacks.
- **Zero-day Attack**: With this type of software-generated incident, attackers can gain access to an endpoint by leveraging a zero-day exploit. Hackers can corrupt or steal data and compromise endpoint functions if successful. In general, zero-day vulnerabilities often lead to zero-day attacks.

Leaders, security experts, and developers must adopt robust security strategies and capabilities that align with their organizations' day-to-day business needs. For example, if endpoint and IoT devices are not appropriately configured, "fragmented identity and unified identity and Access Management may hinder an organization's operational functions and productivity. Therefore, the need for industries to harden their infrastructure protection beyond perimeter defense capability is paramount to daily operational processes. Threats, risks, and vulnerabilities affecting business and individual consumers' endpoint and IoT devices include [30,31]:

- **Malware**
  - Ransomware
  - Trojans
  - Exploit Kits
- **Insecure Network Access**
  - Remote Access
  - Data Transfer
- **Compromised Credentials**
  - Weak Authentication
  - Weak Password Assignment
- **Randomness**
  - Compromised Devices
  - Lack of Endpoint Compliance
  - Insecure Devices in Use

- Poor Device Settings
- Lack of Device Updates
- Unknown Device Access
- Unmanaged Devices
- Zero-Day Exploit

The inability to identify, harden and monitor IoT devices and endpoints is one of the most significant concerns and key drivers for organizations and individual consumers. In addition, the loss of IT productivity and system downtime compromise can negatively impact users' immediate and long-term productivity. Therefore, organizations must invest in advanced security capabilities to protect their IoT devices and endpoints. These solutions prioritize critical systems such as IoT devices and at-risk endpoints requiring continuous monitoring [31].

In ZT, attributes, entitlements, roles, and permissions are developed to protect perimeter-focused network defense cybersecurity systems and perimeterless security environments [32,33].

These security policies, regulations, and controls include [34]:

- **Secure Access Controls**—Organizations must grant enough access using attributes, permissions, rules, and roles.
- **Separation of Duty Enforcement**—allows unauthorized access detection and prevention to thwart data breaches and fraud.
- **Consistent Governance**—Focuses on measuring access effectiveness. Such access includes data, cloud services, and app controls. In addition, these processes are developed for policy compliance and permissions.
- **Deeper Insights**—AI-driven insights provide extensive visibility and insight.
- **Compliance Improvement**—Thorough compliance auditing and maintenance are recommended. Organizations must ensure their employees and users have access to only data and other resources they need to do the job. In addition, their access to those resources must be limited and monitored to prevent data from being compromised.
- **Process Automation**—Organizations must automate and manage their resources. All certificate requests and reviews must be assigned. Managers must develop robust automation processes to support these efforts in their organizations.
- **Identity/Catch Risky Users**—Organizations must monitor users' access and remove suspicious individuals who do not want to comply with the corporate security policy. Such a process ensures that monitored access to resources would help minimize data breaches.

- **Get More Than with Limited Efforts**—ZT's successful implementation means providing low-risk automated access and continuously monitoring high-risk access that might compromise the organization's data and other resources.

Interoperability is essential for implementing ZT policies, managing data/infrastructure, and establishing trust zones. In an organization, the successful implementation of ZT requires everyone's collaboration. Henceforth, the planning and performance of these ZT solutions must be based on cooperation and inter-departmental management's guidance to successfully adopt and deploy ZT resources to secure and protect the enterprise. Ensuring physical and virtual resources is crucial to ZT's long-term success. This process includes the organization's decision to promote collaboration among employees and users. If their policies are clearly defined, leaders and security experts can only define such a roadmap, allocating resources and establishing robust security. How employees and users access organizations' resources and develop security policies to address network segmentation is vital to their long-term business strategy. Organizations represent their network segments and establish user access based on specific roles, permission, authorization, authentication, infrastructure, and geographic locations. In addition, an organization's in-house stakeholder collaboration and operability can establish and encourage all parties to implement ZT core micro-segmentation enforcement and authentication [2,18].

10. Zero trust intrusion detection for IoT systems

Zero Trust (ZT) capability requires Continuous Monitoring (CM) process and applicability. CM is critical for ZT's adoption, implementation, and functioning. Additionally, the ZT environment requires a trust level function between the system, data flow, user, and application. The trust between these critical elements can be dynamically attuned to align with other ZT technology solutions. As changes occur between the system, user dataflow, and application, ZT platforms must also self-adapt for better functioning [23]. Vendors must build automated security solutions to support ZT capability and operational functions. This capability provides autonomous threat detection, isolation, and remediation. Current and future cyber-attacks against IoT platforms can be thwarted if infrastructures are equipped and hardened with proper security capabilities such as firewalls/Intrusion Detection Systems (IDS). IDS are cybersecurity solutions deployed to shield endpoints and networks as passive monitoring devices.

When properly configured and deployed, IDS can detect malicious activities on the network or endpoint. These network/host-based security systems monitor and protect malicious activities in the IoT environment [23,35].

There are two types of IDS systems [23,35]:

- **Host-based IDS**—Host-based IDS (HIDS) protects endpoints. It provides the endpoint with continuous protection against insider, exterior, and other adversarial cyber threats. HIDS has robust security capabilities that protect and monitor "network traffic." In addition, it is designed to provide the endpoint with the solution needed to observe and run processes and examine machine logs. Despite these advanced security solutions, HIDS offers limited visibility for an endpoint. This limitation can hinder decision-makers' abilities. How HIDS provides the host computer with excellent visibility is critical to how the device yields successful results.
- **Network-based IDS**—This IDS focuses on protecting and monitoring the overall IoT computer/network environment. The capability provides 360-degree network visibility and traffic flow. It assesses network threats and determines the best action if deployed to the IoT enterprise. NIDS relies on content and packet metadata to assess, select, and produce results. The device provides a more extensive evaluation of the endpoint or IoT environment in which it is deployed. If properly configured, it detects pervasive cyber threats. Analogous to HIDS, NIDS has limited visibility to any endpoint internals that protects.

Organizations must invest in Unified Threat Management Solutions and cybersecurity capabilities when implementing consolidated security architecture. These solutions are built to enhance HIDS or NIDS detection and protection ability while minimizing the Total Cost of Ownership (TCO). Due to the fifth-generation attacks affecting most global network infrastructures, industries must invest in advanced and robust security to provide real-time protection and offset high TCO. In addition, when deploying IDSs, businesses must invest in improved signature and anomaly detection solutions to protect their endpoint, and IoT works from existing or emerging threats. For instance, "anomaly detection IDS solutions" are model-based systems that we recommend for endpoint and IoT device deployment. This IDS system can identify a system's normal behavior over potential anomalies that can lead to threats. Over the years, security vendors built hybrid detection IDS solutions. This capability provides endpoints and IoT devices with advanced "signature and anomaly-based detection" [35].

11. Internet of things

The genesis of IoT technology gives devices/objects the capability to share data without requiring continuous signature updates or human interference. However, adopting and implementing IoT technology have advantages and disadvantages. IoT is built to provide advanced interconnect solutions and allow physical and virtual devices and objects to interact autonomously as a global infrastructure [36]. This model offers continuous interoperability for communication systems. These autonomous things encompass microprocessors, systems, microcontrollers, and sensor networks. Organizations must develop new security strategies and solutions to address current and growing risks, threats, and vulnerabilities in the IoT environment. These concepts include the introduction of Public Key Infrastructure (PKI) and other selected capabilities, such as the Federal Information Process Standards (FIPS), that organizations may find suitable for their IoT environments [36,37]. Security vendors must rely on PKI solutions to harden their environments. In addition, these solutions offer unique identity solutions that PKI needs to protect IoT systems. In IoT devices, the data flow must continuously be monitored and protected from random attackers. Such protection may include developing a flexible security concept [36,37].

Industry and government leaders and cybersecurity experts must invest in security solutions, that is, digital certificates and authentication capabilities. Recent security studies have proven that PKI's digital/public vital certificates serve as an additional security layer that gives IoT devices/objects the protection they need. Similar studies assert that leaders' and cybersecurity experts' decisions to implement digital public key certificates in PKI can harm their organization's operational and security postures. IPS-140 is a valuable capability for cryptographic modules. FIPS-140 is a de facto best practice and standard for PKI government applications and implementation. As a result, organizations must select FIPS-140 standards when deploying PKI solutions and ensure accurate performance [36,37].

Proper security and protection of IoT systems with continue beyond FIPS and PKI implementation. Therefore, cybersecurity practitioners must approach IoT security from a holistic point of view. Securing IoT devices must encompass planning, configuring, architecting, and implementing all security steps across IoT devices. The following are three notable security layers that IoT systems must adhere to during the implementation and continuous monitoring phases: application, network, and perception layers. The application layer focuses on user interaction and delivers services associated with the user's requirements. Such services include but are not limited to smart home adoption and the implementation of intelligent IoT technologies and solutions [36,37].

In comparison, the network layer is strictly focused on data-centric. In this layer, IoT devices collect data and then process and transmit it to other endpoints. This layer establishes connectivity between IoT devices and other intelligent objects, such as network systems and servers. The network layer generally handles the transmission of data. The perception layer acts as the architecture's physical layer. This layer also allows interconnected devices and sensors to interact and gather large volumes of information in real-time to support single or multiple project requirements. In the IoT cloud environment, the perception layer may act as a sensor, edge device, or actuator which exchanges data with other endpoints within the network. These layers are critical to IoT device security and data protection [36,37].

Cybersecurity experts assert the authentication process to secure and protect IoT devices and objects from current and emerging threats. Authentication solutions have been in use for many decades. The need to authenticate computer and network systems is to establish and verify user credentials and identity. Listed are a few authentication methods that security vendors have developed and are still in use to date: Single Sign-On (SSO), Single-Factor Authentication (SFA), mostly known as primary authentication, Two-Factor Authentication (2FA), and Multi-Factor Authentication (MFA). These security solutions are generally observed in the enterprise's IoT solutions implementation. These introduced solutions, if accurately implemented, can secure, protect, minimize, and protect IoT systems. These advanced capabilities require organizations and security vendors to design and deploy IoT devices embedded with 2FA and MFA functions. These methods provide IoT devices with an extra security layer to stay secure and protected [38].

Manufacturers and security vendors have built IoT systems with biometric security solutions such as touch and face identification and voice recognition. The solutions minimize unauthorized user access, data breaches, network, and resources. However, given the limited functions that 2FA presents to the user when deploying IoT systems, we recommend SSO and MFA. SSO offers users enhanced security functions that exempt them from retaining multiple credential sets. In addition, the processes ensure that users have a unique and perfect experience when initiating the session.

For example, identifying a central domain using an identity and access management system at the organization level ensures that secure SSO links are established. Contrary, SSO allows authorized users to terminate a session and log out from any application or resources they are using [38]. The table below illustrates the difference between MFA, 2FA, and SSO security credentials [39].

In contrast, MFA is still the most preferred and highly guaranteed device authentication method. It uses advanced security functions such as system-irrelevant elements to legitimize user access. These embedded security solutions include device and behavior-based confirmation, password verification, keystroke patterns, and other biometrics functions. In addition, MFA is proven to use multiple-factor authentication. Its ability varies between intervals and combines "elusive elements for invalid users." These solutions offer authentication methods essential to protecting IoT devices, data, and applications. Organizations must sensibly select the correct authentication methods when building and deploying IoT systems. These authentication protocols give IoT systems the security capability to securely operate and protect their applications and data and ensure good usability [34].

Some of these protocols include but are not limited to a Password Authentication Protocol (PAP), Challenge Handshake Authentication Protocol (CHAP), and an Extensible Authentication Protocol (EAP). Despite this notable technological progress, IoT systems are prone to arbitrary insider and external threats. Therefore, security strategies must include how organizations respond to threats and mitigate and recover from insider threats or adversarial attacks. IoT is a system that involves computation, sensing, communication, and actuation. It connects humans, non-human physical objects, and cyber objects, enabling monitoring, automation, and decision making." These computer resources and technology assets are built with key attributes and conditions, allowing interconnected, distributed, and dispersed networks to share data autonomously. IoT spans seven fundamental and interoperable functions: interconnectivity, things-related services, heterogeneity, dynamic changes, enormous scale, safety, and connectivity. In IoT, interconnectivity is a process of connecting intelligent devices and smart objects. This process aims at customer relationship management applications, emails, e-commerce, phones, actuators, sensors, and other interconnected devices or objects [34]. In this context, a function is a set of steps, part of a more extensive computer program.

In contrast, IoT connectivity refers to connecting devices and objects to establish a seamless session over the network. These devices and objects include, but are not limited to, IoT gateways, robots, intelligent lights, and others. Often connectivity is selected after a careful assessment is established and completed. As a result, IoT devices/objects are vulnerable to insider and external threats. Therefore, security vendors must reassess their traditional techniques, methods, and solutions to address risks and manage modern cybersecurity threats targeting IoT systems [34].

Similarly, reassessing current security policies and rules will ensure that organizations and security practitioners have the tools to manage and mitigate current and emerging threats against IoT devices and objects. Thus far, numerous studies cite that organizations and consumers have benefited from adopting IoT solutions. Such benefits include the organization's business growth and consumers' ability to access, collaborate, manage, and share data between interconnected and dispersed devices or objects. Similarly, agility, speed, culture, interoperability, and information sharing are integral components in the IoT environment. However, with these technological advances and massive growth, IoT technology poses threats, risks, and vulnerabilities to organizations and consumers [3,34].

Security vendors note that the following security areas remain a significant challenge in the adoption, implementation, operability, applicability, and continuous monitoring of IoT devices/objects in the enterprise: lack of standards, retrofitted legacy devices that do not comply with the security standards, regulations, and policies aimed vulnerability patches and security updates. These security concerns include but are not limited to unsigned firmware, hardcoded passwords, weak device authentication/access, shared and protected keys, assigned vulnerable encryption protocols, and Distributed Denial of Service (DDoS) attacks. Leaders and security practitioners must prioritize and invest in resources and assets to minimize threats, risks, and vulnerabilities aimed at their organization's infrastructures to manage these constraints. Furthermore, they must ensure that IoT system security must be applied at every stage of the overall organizational-based security model [3].

12. IoT security and monitoring issues

The launch of IoT technology has resulted in large numbers of more portable devices being manufactured and sold in the marketplace. Such consumer demand came with severe mobile security concerns. Due to rising consumer demand, security vendors began building automated products to manage, identify, assess, mitigate, and respond to ever-evolving threats, risks, and vulnerabilities against IoT devices. Security vendors must develop cybersecurity applications that address the rising concerns in the enterprise. These applications are tailored to accessibility, heterogeneousness, and custom-built user needs. These transformed technological threats have helped and encouraged hackers to launch DDoS attacks against IoT devices. Hackers can attack multiple computer resources and technology assets through DDoS by flooding incoming network traffic. These cyberattacks are usually aimed at various IoT devices. After being launched, they are difficult to mitigate [18,27].

Organizations must develop and implement security best practices to ensure continuous platform federations among IoT devices/objects. How these devices collaborate, barter, and provide information is vital to business success and stakeholders' needs. Interoperability is a capability that allows for device data exchange at the enterprise level despite organizational, political, and social factors that may impact machines' performance. Leaders and employees can communicate if policies are unambiguously defined and implemented as practical communication tools. For instance, information disseminated within an organization explains the brand and culture leaders often embrace within their institutions. The need for better collaboration in an organization can exemplify excellent organizational interoperability. This concept can be applied to IoT devices, objects, and user interoperability. The heterogeneous concept sometimes involves wide-ranging areas, such as application, data, semantics, network, middleware, and devices [18].

The need for standardized interoperability in the IoT environment remains the most significant factor for leaders and technical teams. IoT devices/objects continue to experience significant security and interoperability issues. Such challenges are due to the extreme heterogeneity between IoT systems and objects. This heterogeneous environment includes device data formatting, communication protocols, standards, and other technological solutions. For instance, the need for unified standardization of IoT device implementation/deployment remains a more significant challenge to developers and organizations that intend to implement such technologies.

Consequently, "Interoperability" was initially developed to address the continuous interaction organizations need when developing products and systems. The term spans information technology and systems engineering domains/services. Clear communication and meaningful data/information transfer are vital to an organization's success. Optimal operational interoperability encompasses various information systems and how the data is disseminated or provisioned between IoT devices and objects [3,18].

The more intelligent devices and objects are being deployed into the highly fragmented IoT-based environments, the four types of interoperability, namely, organizational, technical, semantic, and syntactical, are critical to the systems' operational, provisioning, and interaction. At the corporate level, interoperability is essential for deploying green IoT devices. This process may entail deploying enhanced cross-domain applications to secure and protect day-to-day business resources and operating assets. The method includes the interoperability's need for an abstraction layer that allows for the smooth deployment of cross-platform applications [18,27,36].

In IoT environments, heterogeneity allows devices to interact and share data efficiently with balanced device service consumption. Vendors have developed the User Interoperability Framework (UIF) to allow device users to interoperate [27]. With the UIF device, users can interact and share information with their peers/other users who possess heterogeneous devices using different contexts, word order, and semantics. The UIF prototype is introduced and implemented through experimental approaches [18]. It focuses on ascertaining the best way to adopt semantic relatedness computer implements. Interoperability's identification and resource components allow vendors and developers to build new IoT devices/platforms embedded with advanced sensory technologies. The need for organizations to adopt interoperability when adopting and deploying their IoT systems guarantees system interconnectivity, data provision, and interaction [36].

There are three types of IoT interoperability [36,40]:

- **Technical Interoperability**: Provides organizations the capability to deploy physical communications infrastructure to provide chunks of information/data sets. This type of interoperability allows for seamless data exchange between IoT devices. As more IoT devices and objects can be adopted and deployed into the ecosystem, the need for enhanced interoperability between things, users, and data is key to a healthy operational environment. In IoT, environment interoperability is how single or multiple platforms access, share, and provision data across platforms. Such a process requires multilayers of Interoperability between IoT systems. When building or transitioning from on-premises and virtual IoT environments, leaders and their technical teams must consider these five pillars of interoperability: People (User) Trust, Device Trust, Network Trust, Application Trust, and Data Trust. Organizations must incorporate these five ZT pillars when building the ZTA. For instance, technical interoperability allows for machine-to-machine data exchange. Therefore, these IoT systems can be interconnected within the same physical or virtual environment.

Leaders and technical teams must define adequate standards and protocols for interoperability and system performance.

- **Semantic Interoperability**: How semantic interoperability functions are based on deploying IoT systems. It supports IoT deployments based on establishing the data meaning. Such events include but are not limited to how IoT devices communicate and exchange information effortlessly. Despite its name, leaders and technical teams often define their protocols for semantic interoperability. Semantic interoperability can be applied using a variety of approaches. For instance, it can be applied in the Big Data for IoT, known mainly as SIMB-IoT. This interactive function aims at delivering semantic interoperability for heterogeneous devices. A rigorous approach must deal with the complexity of semantic interoperability. Organizations must set forth proper standards to support their IoT environment in semantic interoperability. These guidelines include but are not limited to mapping data and frameworks. The structures define the information model and type of IoT platforms an organization requires to function correctly and maintain adequate operational posture.
- **Syntactic Interoperability**: This approach is based on shared syntax or "common information model structures." The concept is based on collective word order, aka syntax. Generally, syntactic interoperability provides data structures and determines which protocol can communicate the information. In a syntactic environment, semantic IoT deployments must establish data significance. Specific typed data is vital in establishing and structuring data format before sharing.

If industries lack interoperability, maximizing productivity and value can be difficult while demonstrably impacting long-term innovation in emerging technologies such as predictive automation, artificial intelligence, machine/deep learning, digital twins, etc. Interoperability is a selected concept that allows IoT devices to communicate, share data, interact, and provision information physically and heterogeneously [27].

As these IoT platforms are developed, they are assigned a device identification. In essence, all devices/objects are assigned an identification system. In IoT, the device identifier is based on resource request format though there are variations between systems. Despite a positive process between interoperability and heterogeneity, most vendors and developers need to pay more attention to the device identification systems. In recent years, some IoT vendors and developers have considered building "things" embedded with Device Name Systems (DNS). This architecture-based concept focused on heterogeneous comparative/relative analysis [27,36].

Yesteryear, security experts and vendors built applications to protect the organization's physical and virtual resources. Organizations must continue exploring methods, techniques, and solutions to protect their perimeter-based cybersecurity systems. These concepts are vital to organizations' daily operations and customers' business needs.

These concepts, ideas, and methods involve a triad approach [33]:

- Device security and provisioning.
- Establish secure endpoint and cloud connectivity.
- Specific data in transit and stored in the cloud.

Perimeter-based defense cybersecurity solutions such as virtual local areas and private networks have secured computer resources and technology assets for many years. However, this network security configuration has shifted beyond a perimeter-focused defense cybersecurity system. Virtual local area networks, network access control lists, and Virtual Private Networks (VPN) were considered an extra security model in a traditional computer and network security environments. VPNs were built to protect perimeter-focused defense computer resources and technology assets from external intrusions. Aside from these notable technological and scientific advances, security vendors reevaluate perimeter-based defense postures, controls, methods, techniques, and policies [33].

In today's ever-changing cybersecurity environments, network services and managed endpoints are exposed to internal and external cyber threats, risks, and data breaches. As a result, organizations view the need to invest in state-of-the-art cybersecurity applications as critical to their internal security and address clients' needs. Because of that, security vendors, leaders, and cybersecurity experts must explore robust solutions to secure and protect physical and virtual resources [33].

Standardized policies and solutions are needed to help minimize and mitigate current and future threats, risks, and vulnerabilities. As organizations explore a way to implement these solutions, they have to inventory their existing endpoints within and beyond the perimeter-focused defensive perimeter. For instance, security experts envision Permanent Denial-of-Service (PDoS) attacks, generally known as plashing, as detrimental to IoT devices. DDoS and PDoS attacks have escalated in recent years. As part of their advanced and long-term strategy, hackers often rely on malware such as BrickerBot to launch random attacks on IoT devices [18,33].

Therefore, establishing controls beyond security pillars is critical for an effective ZT implementation. This approach allows organizations to employ integrated strategies and implement security solutions to secure the infrastructure.

Attackers use a type of malware, BrickerBot, to manipulate specifications and passwords on IoT devices. Once they have successfully cracked the device passwords and specifications, they could launch PDoS against one or multiple IoT systems. The primary intent is the PDoS attacks is to disable passwords on IoT devices. Attackers use various security methods and techniques, Cross-Site Scripting (XSS), Cross-Site Reference Forgery, and CVE-2016-8581, to indiscriminate attacks on IoT devices. Cybersecurity applications/malware, e.g., Mirai and BrickerBot, can cause long-term damage to IoT devices. By and large, hackers can launch random attacks on their IoT devices through these sophisticated web tools. They are using Structured SQL-Query Language injection web interface attacks. Hackers use XSS to explore vulnerabilities in Drupal 8.X. This process allows attackers to randomly send web scripts through an HTML, mostly known as HyperText Markup Language file permeating trajectories with HTTP-Hypertext Transfer Protocol medium [2,18].

As part of a long-term security and exploitation strategy, hackers often select the data integrity attack by manipulating information to gain 360-degree control of IoT systems. Accordingly, security vendors must develop cybersecurity applications to counter existing and emerging threats, risks, and vulnerabilities. These experts promise enhanced security protocols and mechanisms such as machine-to-machine encryption. These applications span Elliptic Curve Digital Signature Algorithm, Secure Hash Algorithm, and Hash-based Message Authenticated Code. Embedding these cryptographies with asymmetric/symmetric keys can help establish a two-way authentication between IoT devices. These advanced security protocols communicate between distributed or dispersed IoT devices and the cloud. In addition, they protect data transmission sessions between users and IoT machines [18].

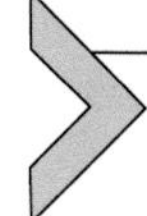

13. Data privacy, policies, and regulations for IoT devices

Data privacy, policies, and regulations are critical to IoT operations and information flow. Excellent policies and regulations are needed to ensure that IoT objects function within proper safeguards. How IoT devices are used comes with enormous benefits to consumers and disadvantages if illicitly used to cause harm to organizations and individual citizens' rights to data privacy. Lack of proper privacy may allow hackers to access organizations' and consumers' data illicitly. Studies have proven that consumer confidence in IoT devices and other technologies can be undermined if consumers' privacy issues are not addressed. Better regulations and policies must be instituted to protect organizations' and consumers' useability and accessibility of data, whether via the Internet or communication communications. Organizations must introduce security processes, policies, and guidelines to assess, thwart, mitigate, or eliminate cyber threats against IoT systems [41].

Over the years, most businesses have invested and developed models, architectures, and frameworks to protect their IoT infrastructures and mitigate private-based risks affecting their daily operations. These tools can help minimize data and safeguard the information against access to unauthorized users. In addition, assessing data practices and business requirements or developing policies that guide and articulate such practices help establish confidence and limits how data is collected, shared, and retained. However, the increased demand for converging IoT objects, data, and how these devices connect to the Internet must be clearly defined and adequate policies instituted. Therefore, these guidelines must focus on data convergence, device connectivity, data economics, and the next generation of information flow and device interface. In addition, any policies being instituted must further address the IoT ecosystem, architectures, and frameworks that govern the development and deployment of these systems. Accordingly, such proven guidelines must support the robustness of IoT platform environments and develop enhanced security capabilities [41].

14. IoT device hardening and security monitoring

Organizations must invest in security and monitoring capabilities to protect their IoT systems. Security vendors must develop advanced applications, i.e., Transport Layer Security (TLS) and Secure Socket Shell (SSH). TLS and SSH must be used as alternative solutions for securing IoT applications and their data flow using HTTP and HTTP/2 protocols. We recommend the SocketXP solution for remote users accessing their business resources. The solution provides organizations and users with seamless device management. It allows authorized users to access the organization's resources remotely. SocketXP offers SSH a remote capability to establish a session with IoT systems. Once a session is established, users can initiate a secure tunnel with firewall(s) or network access translation router(s). The session is established via the Internet to develop SSH access remotely. Over the years, most IoT devices were installed without adhering to proper system configuration and security implementation features/standards. Such negligence has yielded widespread system vulnerabilities and incidents [42].

Lacking authentication protocols to harden and secure the environment can pose immense threats, risks, and vulnerabilities to IoT systems. Security vendors noted that at least 98% of IoT system traffic is unsecured [43]. The lack of standardized security protocols has exposed organizations' and consumers' data to hackers. Most attacks against IoT systems are sophisticated. Hackers use these unencrypted to circumvent perimeter defense networks using phishing attacks to gain 360-degree control of the enterprise [42,43].

15. Cyber risk management for IoT device security

IoT security focuses on the organization and individual consumers' protection of IoT resources and related technology assets. This concept is developed to ensure that industries and US Government agencies have the tools to assess and mitigate threats, risks, and vulnerabilities. In addition, improved cybersecurity frameworks and tools are designed to streamline the time and resources allocated to perform IoT platforms' cybersecurity risk assessments and management frameworks. Due to the lack of traditional "security protection technologies," vendors developed newer cyber risk management frameworks and tools to tackle the ever-growing data breaches and malicious cyber-attacks on IoT platforms. Organizations must invest in better frameworks, architectures, and embedded devices to address and mitigate emerging cyber threats, risks, and vulnerabilities. Such efforts often require the involvement of multiple entities, for example, the organization's internal management, vendors, and the customer [44].

IoT, a cyber risk management framework, entails these four layers [10,44–47]:

- **IoT Cyber Ecosystem Layer**: This layer offers stakeholders and IoT systems the ability to interact and compete. These stakeholders range from "IoT cybersecurity and technology developers." Other stakeholders, such as governments, standardization organizations, adversaries, customers, and external users, often benefit from the IoT cyber ecosystem. Immediate attention from cybersecurity managers allows for IoT cyber ecosystem changes. Such swift changes are based on the IoT systems' proper formulation of security response. Many IoT managers have yet to mature their expertise in IoT cybersecurity solutions. To ensure IoT managers have competitive technical skills to perform on the job, they must continue to leverage their IoT cybersecurity knowledge. The industry must train more IoT cybersecurity technology developers despite continued growth in the IoT marketplace. To improve IoT device specifications through innovation, IoT cybersecurity developers must leverage their solutions which are 5G, edge or fog computing, IoT systems, sensors, serverless, and other related technologies. In addition, IoT systems must be equipped with robust and higher accuracy capability measures for detecting, protecting, and recovering if they are compromised. To ensure better protection, detection, and recovery mechanisms, IoT cybersecurity developers must evaluate traditional methods before developing end-to-end IoT solutions that can adequately address and deliver seamless security capabilities to protect and defend IoT security platforms from insider and external intruders. For better solution deliverance, some level of communication between the user and customer must be established to ensure the protection, detection, and delivery of robust IoT solutions.

- **IoT Cyber Infrastructure Layer**: Organizations must develop and assess their future-oriented cyber risk management plan. This process includes developing and evaluating IoT cyber infrastructure from technological and managerial perspectives. The ability to protect IT assets and/or services involves risk management technical aspects. These measures are developed to safeguard IT services and assets. If these safeguards are not adequately developed and implemented, they can be detrimental to organizations and their users. Such IoT cyberinfrastructure activities include, but are not limited to, implementation, monitoring & control, and continuous improvement. Each of these activities is integral to the functionality of the IoT cyber infrastructure layer. Organizations must develop cyber risk management solutions to address current and future threats. These solutions include technological and managerial perspective evaluation of existing IoT cyberinfrastructure. Protecting IoT assets and services must be the organization's priority when planning and developing future cyber risk management solutions. Employees and other organizations' internal users adhere to the IoT cyber infrastructure capability. The IoT cyber infrastructure layer spans behavioral system elements such as those factors affecting cybersecurity technology, security, and privacy policy compliance. Organizations must promote cybersecurity awareness and develop best practices to support the management of cybersecurity technologies. Every organization needs robust cybersecurity policies to protect and defend its IoT systems. Senior management must establish guiding principles to develop cutting-edge cyber capabilities to identify, protect, and neutralize emerging threats against their IoT assets. Therefore, top-down management support within an organization is key to developing and implementing agile security solutions. Such capabilities must align with the organizational cyber operations strategies and governance designed to address ever-changing attacks and threats. For instance, the relevance of cyber technology assets might determine how IoT cybersecurity technology developers must plan and develop robust IoT cyber solutions to assess, protect and recover from current and future threats. Some of these IoT cybersecurity technologies include but are not limited to intrusion detection and intrusion protection systems. Other security solutions include "encryption, public key infrastructure, tokenization, firmware, device authentication, security analytics, firewalls, Distributed Denial of Service, incidence response systems, secure communication, and device authentication, etc." [10,45–47]

- **IoT Cyber Risk Assessment Layer**: Organizations must identify and assess their current and future cyberinfrastructure. Organizations must follow these processes by assessing their cyber risks: identify and quantify risk and allocate resources to perform such tasks. When identifying IoT assets, vulnerabilities, and cyber threats, the high-level process must include how intruders can prepare and launch random cybersecurity attacks against IoT assets. Most cyber-attacks can be launched using explorative or exploitative methods and techniques. As a preventive measure organization against these explorative and exploitative cyber-attacks, we recommend Cyber Kill Chain (CKC) model and/or framework. CKC tool identifies and prevents risks intruders may launch against the IoT assets. Lockheed Martin developed the CKC model and/or tool as part of its internal Intelligence Driven Defense capability. The tool identifies and prevents cyber intrusion threats that intruders, such as internal and external actors, can launch against IoT assets. In general, intruders can launch traditional attacks against IoT systems variety of methods, such as phishing flight risks and persistent insiders. Disgruntled employees often launch flight risk attacks with malicious intent within the company. Disgruntled employees generally have a tendency randomly launch these less sophisticated attacks, which lead to an organization's loss of data. At the same time, persistent insider attacks are more sophisticated attacks that can be launched internally. Most employees who launch these types of insider attacks do not want to separate from the organization. Instead, the employees are interested in causing sabotage by accessing the organization's sensitive data to gain self-profit. Most employees have extensive cybersecurity knowledge and can access the organization's assets by surreptitiously bypassing security controls and internal data. Some employees commonly attempt to gain access to organizations' data and systems that permission cannot be granted to them. Risk qualification can be defined as a three-prone method spanning stage, impact, frequencies, and defense. Most of these methods and techniques can be used to measurably quantify cyber risks against IoT assets by grouping several cyber threats into one quantifiable measurability. Cyber attacks' frequency can be measured by how often the attacks can be launched against single or multiple IoT assets. Using different metrics, cyber experts can determine the impact that such attacks can have on IoT assets. The outcome of these measures could ensure that cyber experts clearly understand the attacks launched against IoT assets, some of which may result in data records thefts and financial losses. Cyber experts must develop a risk matrix to mitigate and understand the vulnerability, risks, and threats that may affect IoT assets and the organization's day-to-day operations.

- **Cyber IoT Performance Layer**: The Cyber IoT performance layer encompasses the implementation, monitoring & control, and continuous improvement process. Cyber experts are responsible for identifying cyber solutions and allocating resources needed to support the performance activities. After identifying and allocating resources in the previously discussed IoT cyber assessment, cyber experts generally determine whether the performance activities stage is warranted. The cyber performance layer is a necessary process for conducting risk management. Cyber experts must make decisions based on findings garnered from the risk assessment layer. Once a decision is made, these key activities: "implementation, monitoring & control, and continuous improvement," must be reviewed to ensure that each of these processes aligns with the IoT cyber performance layer's requirements. The implementation phase encompasses "development, testing, deployment, new policy development, training, and user acceptance." Cyber experts must develop robust evaluation capabilities that will allow them to select preferable criteria from a host of cyber technologies that could be identified from the IoT cyber ecosystem layer. When implementing such solutions, the following process criteria must be considered: cyber monitoring & control systems' usability and usefulness. In theory, most commercially identified cybersecurity platforms can be procured and deployed to assist with the monitoring & control systems development to support the customization of the solution being offered. In the monitoring and control stage, cyber experts must also consider different types of cyber activities that can be taken into consideration—prevention, recovery, and detection. This stage is designed to monitor and facilitate the control process in response to known or emerging cyber-attacks against IoT assets. Both prevention and detection activities focus on data collection of abnormal users and equipment performance ranging from data and applications to illegal access to unauthorized IoT systems. In addition, cyber experts must have a recovery plan to counter or thwart any malicious activities against designated IoT assets if a data breach is identified. This process can be carried on by delivering a capability needed to mitigate such attack(s) in real-time [10,47].

IoT platforms must be certified before and after deploying into the environment. The IoT cybersecurity risk management architectures and frameworks comprise the perception, network, processing, application, and service management layers [44]. This set of architectures and frameworks encompass the NIST, ISO/IEC27005, and other cybersecurity frameworks, architectures, and methods such as the CMMI—Capability Maturity Model Integration, Bayesian Decision Network, AVARCIBER, CKC, and OCTAVE—Operationally Critical Threat, Asset, and Vulnerability Evaluation that can be applied to the IoT device environment and beyond [44].

16. Augmenting security in embedded IoT platforms

Cybersecurity threats are evolving and remain the most concern for the industry, government leaders, and private citizens. Continuing data breaches and cyber-attacks on multiple IoT infrastructures can be categorized in three-dimensional sequences: Confidence, theft of services, availability, and data integrity. These security threats to IoT platforms pose the most significant concern to industries, governments, and consumers. In addition, most cyber-attacks on IoT devices can be launched from individual or dispersedly wired or wireless network systems. For that reason, these devices/platforms can be dispersedly located throughout the world.

Therefore, the need for enhanced embedded solutions can help minimize current and future cyber threats on IoT platforms. The following are security methods we selected for embedded IoT security solutions [48]:

- **Message Replay Protection**—This method facilitates packet encryption by enhancing such a process with data fields. Using this technique, recipients can implement sequential acceptance rules. Such policies can be instituted to prevent unauthorized users from accessing recorded messages. Unfortunately, the restrictions do not apply to decrypted messages being sent again, which may trigger a security incident.
- **Message Authentication Code**—This code is typically executed by "cipher/hash algorithms." Hash algorithms using data packets often create a short signature. Once the message is sent, the receiving party uses the exact mechanism to verify the sender and ensure that the received message has not been altered. This process ensures that encrypted messages are not intercepted and compromised while transmitting between the sender and the recipient. Despite being a "low-complexity method," this technique can be implemented correctly in embedded IoT devices.
- **Debug Port Protection**—Most embedded IoT devices use port protection to configure, analyze, and control the debugging process. A factory password is required to protect IoT devices during the debug process. Once purchased and deployed as field units, the factory password in those IoT systems can be disabled internally to protect the machines from random cyber-attacks. For instance, the "firmware development and debugging" process can be achieved by using the Joint Test Access Group, aka "JTAG/Serial Logging" port(s).
- **Secure Bootloader**—This firmware can be built-in and quickly loaded into embedded IoT devices to restrict unauthorized users' access. Secure Bootloader (SB) is built using cryptographic hash algorithms numerally assigned with the vendor's private key.
- **Pre-Shared Keys**—IoT systems are embedded with matching Pre-Shared keys (PSKs). Because of their limited security capabilities, if implemented, we recommend that PSKs be coupled with digital certificates to provide a robust level of security for IoT devices. These security keys are usually stored in cloud-based servers. Due to their limited security capabilities, organizations must invest in advanced security solutions that can be used as a second security protection layer to PSKs. Although PSKs can be vulnerable to adversarial random security and data breaches. Most attacks on PSKs are launched over the air,

via man-in-the-middle, and through brute force or an air crack-ng tool. In addition, hackers often launch attacks on IoT systems through a simple dictionary, many types of layer 2, VPN, Firewalls, password theft or loss, and phishing vectors. Security vendors must develop enhanced tools to protect IoT devices and other embedded platforms.

- **Secure Shell**—Secure Shell (SSH) is a protocol to protect configuration and debugging ports. SSH is used to establish encrypted console connectivity. In addition, the protocol is used for port password debugging. Despite its advanced capability, SSH implementation in embedded systems can be complex. In the larger Operating System-based, implementing this protocol can be relatively quicker and smoother.
- **Public Key Exchange**—Public Key Exchange (PKE) is a protocol that provides a one-way function capability. If a hacker intercepts the PKE message during the protocol execution, it may be challenging to compute the "secret keys." Such inability would prevent the attackers from accessing a shared "secret key k." PKE is often used as a cryptographic method in the IoT environment, allowing for the critical establishment or exchange between two parties.
- **Transport Layer Security**—Transport Layer Security (TLS) is a proven standard that encompasses a broader functioning of the Secure Socket Layer. TLS is the universal model for PKE. It provides an encryption protocol for securing message trafficking between IoT systems. TLS can be executed on embedded systems such as Linux-based machines. Such embedded systems may support IP sessions running Transmission Control Protocol. Such a concept may be applied in small-scale embedded systems supporting TLS. Before making such a decision, organizations must assess the need and allocate appropriate resources to support such an effort.
- **Wi-Fi Protected Access**—Wi-Fi Protected Access 2 (WPA2) can be used to secure embedded computing devices so that their sessions can be initiated and established via Wi-Fi (802.11 ac/ax). Furthermore, the protocol allows seamless interoperability between embedded systems after an established connection via Wi-Fi. Therefore, we recommend WPA2 as an alternative solution for large-scale embedded systems. Furthermore, WPA and WPA2 can be deployed as a suite of security protocols to protect secure wireless networks. The Wi-Fi Alliance developed this certification program to address several security vulnerabilities that experts identified in embedded systems requiring the Wireless Encryption Protocol or Wired Equivalent Protocol.

- **Packet Encryption**—focuses on data exchange and provides an extra security protection layer to the IoT systems. In addition, it contains embedded computing devices that provide additional protection while allowing for the implementation of essential encryption solutions. Some of those encrypted solutions include the FIPS 197 and AES solutions. These capabilities can be implemented to protect message trafficking from being accessed or compromised by unauthorized users. Besides, message replay protection can easily be deployed to enhance IoT device protection when used with private keys.

Organizations must develop and implement advanced strategic planning and tactical analyses, robust security controls, and applications such as add-on security solutions to tackle cyber threats, risks, and vulnerabilities affecting their IT infrastructures. Most industries that have implemented IoT platforms rely on these two monitoring and mitigation methods:

- **Failure, Models, and Effects Analysis (FMEA)**—Is a standard core technique organizations use to analyze their IoT environments. We recommend FMEA as the security analysis for assessing risks, the level of gravity involved, probability, and detection. FMEA gives security experts the granularity to pinpoint the level of threat involved and mitigate it. These risks are identified based on a priority scorecard that defines and determines the level of risk that requires the most attention for remediation.
- **Damage Potential, Reproductivity, Exploitability, Affected Users, and Discoverability (DREAD)** – This is another unique assessment method we recommend for organizations to employ as part of their security analysis approach within their IoT environments. As security experts identify risks, five impacted categories must be carefully calculated and tallied. DREAD relies on the resulting risk score to identify and prioritize how to mitigate and respond to such adverse threat levels.

For better results, security methods for all embedded IoT systems must be carefully identified, selected, and employed. Employing security systems on traditional machines differs from how similar solutions can be deployed in an IoT environment. Embedded IoT systems present several security constraints that traditional platforms do not.

Some of these constraints include but are not limited to [48]:

- **Interface Differences**—How embedded IoT systems interface when deployed in the environment differs. Such differences may include the device specifications and connectivity that most platforms present.

- **Limited Resources**—Embedded IoT objects do not have extended battery life, good processing speed, and memory space.
- **Basic User Interfaces**—These devices do not have an embedded graphical user interface. Instead, their error messages are based on flashing lights and beep code series. Unfortunately, the same applies to security status/command events.
- **Hardware Ports**—Are prone to exploitation and malicious viruses.

Aside from their reliance on wireless networks, embedded systems are exposed to cyber-attacks. Besides, the "new wireless links" have triggered the physical access barrier removal. Unfortunately, these setbacks have encouraged hackers to increase cyber-attacks on IoT devices and platforms [48].

17. AWS and Azure IoT solutions

Amazon and Microsoft are two of the leading global IoT cloud-based service providers. AWS has developed more IoT solutions despite providing higher services than its competitors, i.e., Azure and Google. The company builds and deploys solutions for cloud services and edge software. Through edge software, developers can establish device-to-device connectivity. Edge software allows developers and machines to collect, process, and make informed/intelligent decisions regardless of limited Internet connectivity. AWS's approach to securing IoT follows the recommendations of NIST's Zero Trust Architecture. The goal is to provide a ZT framework for every deployed endpoint of their cloud services, using identifiable and certified devices with manageable software and firmware. In addition, AWS and its partners have implemented IoT ExpressLink to create validated hardware and software for a secure machine connected to their cloud. They include a suite of services for monitoring and defending their compliant devices. As a result, cloud services provide device-to-device secure connectivity. In addition, these services give the machines security detection and response capabilities [20,49,50].

AWS, Azure, and Google have built security applications, products, and services to deter lateral movement or vector attacks against IoT devices and objects. Such services and solutions include but are not limited to Edge Computing (EC). EC is a cloud layer that brings cloud services and applications nearer to the data generation source in the AWS IoT environment. Azure and Google platforms have similar functions despite their strategic and tactical capabilities that differ when implemented in the enterprise.

For instance, Amazon built edge software to provide integrated cloud solutions to support data management and processing. AWS edge computing has two types of IoT solutions: "FreeRTOS—Real-time Operating System and AWS IoT Greengrass." Amazon FreeRTOS is an integrated and interoperable open-source Kernel solution designed for micro-controllers. It provides software libraries that AWS needs to establish a secure connection between small and low-power IoT devices to the AWS cloud services, explicitly edge and core devices operating "AWS IoT Greengrass." AWS IoT Greengrass provides devices with the extra capability to steward the information they process within the cloud environment. It uses the cloud environment for analytics, management, and secure storage. In AWS IoT Greengrass, devices need "AWS Lambada" to perform various computational tasks. For example, these devices can make predictions using machine learning simulations. In addition, they can transmit and process data with or without Internet connectivity. AWS IoT cloud services comprise IoT machines with embedded applications [49,51].

AWS IoT cloud services include [51]:

- AWS IoT Core
- AWS IoT Device Management
- AWS IoT Device Defender
- AWS IoT Analytics
- AWS IoT SiteWise
- AWS Partner Device Catalog
- AWS IoT Events
- AWS IoT Things Graph

AWS IoT infrastructure implements the NIST model, specifically [32]:

- All data sources and computing services are resources.
- All communication is secured, regardless of network location.
- Access to individual enterprise resources is granted per-session basis, and Trust is evaluated using the least privileges before access is granted.
- Access to resources is determined by a dynamic policy that includes the observable state of client identity, application, service, and requesting assets, all of which may consist of other behavioral and environmental attributes.
- No asset is inherently trusted—the enterprise monitors and measures the integrity and security posture of all owned and associated assets. In addition, the enterprise evaluates the security posture of the asset when evaluating a resource request. Therefore, an enterprise implementing a ZTA must establish a nearly continuous diagnostics and mitigation system to monitor, patch, and fix the state of devices and applications.

- All resource authentications and authorizations are dynamic and strictly enforced before access is allowed. This concept involves a continuous cycle of obtaining access, scanning and assessing threats, adapting to threats, and reevaluating the Trust in ongoing communications.
- The enterprise collects as much information as possible about the current state of assets, network infrastructure, and communications, which it uses to improve its security posture.

While this approach, architecture, and services provide practical and manageable IoT security, it is somewhat limited by using AWS-certified devices. As a result, it cannot fully address legacy IoT device deployments.

Azure IoT is a compilation of cloud services that can be adopted and deployed to establish a connection and manage billions of IoT assets. In addition, these cloud services monitor IoT assets. Azure IoT cloud services are cheaper than AWS/Google IoT solutions in service offerings.

The Azure cloud comprises [51]:

- Platform as a Service (PaaS)
- Azure Digital Twins
- Azure IoT Central
- Software as a Service
- Azure IoT Cloud Services
- IoT Hub
- IoT Edge
- IoT Hub Device Provisioning Service
- Time Series Insights
- Azure Maps

17.1 Azure IoT solutions accelerators

These collected works of PaaS solutions are built to speed up the development of Azure IoT solutions. Azure services can be tailored to IoT solutions to meet clients' requirements. These requirements would require engineers and developers with .Net and Java skillsets. Such expertise is often needed to support the JavaScript back-end day-to-day operational activities. These models allow for the device, user, and space collaboration and interaction.

17.2 Azure digital twins

This Azure capability allows the development of across-the-board standards that support the physical environment in the IoT environment.

17.3 Azure IoT central

This Azure SaaS capability allows IoT devices to connect, manage and monitor performance and activities within the environment. This fundamental application is a capability that the Azure technical team would use for machine-to-machine monitoring and provisioning. The service can be deployed as a fundamental software solution. The central application does need a deep-dive level of customization. The Azure team does not require advanced programming skills to provide IoT system configuration.

17.4 Azure IoT cloud services

Establishes connectivity between machines and the Azure cloud. This connectivity allows for data analysis and processing.

17.5 IoT hub

The capability allows the Azure Technical Team to establish connectivity between IoT systems and the IoT Hub. IoT Hub focuses on IoT device monitoring and controlling. This solution is suitable for bio-directional IoT systems and back-end communication.

17.6 IoT hub device provisioning service

Allows users to provision machines securely to the IoT hub. In addition, IoT Provisioning enables users to provision several devices swiftly. This process allows for the provisioning of multiple devices rapidly.

17.7 IoT edge consists of three Azure components

IoT Edge Modules, IoT Edge Runtime, and the Cloud-based Interface. These components can be containers running Azure modules. Edge is built to perform "Cloud Analytics and Business Logic." Moving the workload to Edge could minimize the message trafficking between local devices or on-premises environments.

17.8 Time series insights

Built to collect, visualize, request, and process large time-series data sets. These services generate extensive time-series data for processing. This solution is available to support the IoT Hub applications and services.

17.9 Azure maps

These are Azure services that deliver "geographic information." These services support mobile and web application environments—a set of Web-based JavaScript Control and REST APIs power Azure Maps. Web-based JavaScript Control and REST—REpresentational State Transfer APIs are designed to build flexible applications.

18. Google cloud IoT solutions

Google is one of the leading cloud providers in the industry. Google developed the cloud platform to provide public and commercial clients with wide-ranging services as part of its technology brand [52]. Google's cloud vision included the WireQueue Message Queuing Telemetry Transport (MQTT) Toolkit. Google uses WireFlow's MQTT to provide additional layers of IoT devices. The toolkit serves as a protocol for IoT-based systems to exchange data with/through the Google Cloud Platform (GCP). The MQTT protocol acts as a conduit to transport data from IoT devices to the cloud once the session is initiated. Google did not develop its standard MQTT broker to work in place of the GCP.

GCP is a cloud-based server application deployed using the MQTT protocol as an Internet-based Infrastructure Application. The protocol establishes and enforces operation rules/guidelines [52]. When deployed, IoT devices send sensory data to the Google cloud. A JSON Web Token (JWT) initiates a session and establish connectivity with GCP. A JWT guarantees secure connectivity between IoT devices and the Google Cloud environment in GCP. Google uses Python code to generate JWT sessions. IoT devices use telemetry to measure and collect data. Telemetry measures data collected from remote or inaccessible IoT endpoints as an automated device. Once it has collected the data, it sends it to other devices for monitoring [52]. The Google Cloud IoT speeds up operational agility and time provisioning data from IoT devices to the cloud. The platform provides real-time automation, analytics, continuous monitoring, and intelligence insights between edge computing and the cloud environment. Through its BigQuery, Vertex AI, and Google Data Studio, Google Cloud IoT provides Ad-Hoc analysis, data visualization, and advanced analytics capabilities. In addition, the cloud offers enhanced operational efficiency, real-time IoT data intelligence, and additional tools that IoT devices need to function autonomously [52].

Google Cloud comprises the following unified building blocks [52]:

- DataFlow
- Pub/Sub
- IoT Core
- VertexAI
- BigQuery
- Additional Google Cloud Services

Each of these cloud solutions aims at providing advanced capabilities that organizations and consumers need to stay operational. For example, Google Cloud allows end-users to manage IoT data networks from a centralized location. Google's IoT core is built to interact with Commercial Off-The-Shelf products.

The lack of IoT device capabilities begins with the vendors that manufacture them. For instance, the design of IoT security solutions for basic Radio Frequency Identification/RFID and programmable logic controllers is generally an afterthought or completely ignored. As a result, such devices' significant and widespread vulnerability consists of initial programming, setup, default parameters, and passwords [53]. Consequently, many network technicians, installers, or end-users of IoT devices often do not update embedded device software and security policies. Such inability is due to a lack of knowledge, time, documentation, interoperability, and other capabilities. On the whole, hackers focus on these gaps to exploit vulnerabilities in the devices.

In most cases, attackers target IoT devices, i.e., home security cameras, network-connected printers, and even "Smart Light Bulbs" [54]. As noted, recommendations, guidelines, and standards are left unimplemented to eliminate or minimize such risks. Commercial and consumer IoT devices can connect to existing networks through IP services, e.g., Wi-Fi or proximity links like Bluetooth, direct connection using Ethernet, and cellular phone interfaces. Once connected to a device, hackers or sophisticated automated botnets can use a back door to gain full access to the local area network. Such control includes but is not limited to altering device specifications, monitoring network traffic contents, and potentially accessing privileged links to other services [55–57].

Gaining initial access to IoT device security specifications can be trivial. Yet, cybersecurity experts must periodically monitor these types of breaches. Hackers can achieve such tasks by searching online device documentation, among other essential reference materials. Aside from these concerns, security experts remain most vigilant about the threats, risks, and vulnerabilities targeting organizations' IoT systems. For example, a recent law in California [58] requires "all Internet of Things (IoT) devices sold or purchased within the state be equipped with reasonable security measures." A similar law applies to Oregon, where companies manufacturing these IoT devices, such as in California, must comply, or else they could receive legal penalties [53,58].

In the United Kingdom, a recently proposed "IoT Security Enforcement Body" may "...have the power to ban, recall or destroy insecure consumer IoT products...and could be "...granted the power to apply for a court order to confiscate or destroy a dangerous product or issue fines against the manufacturer "[59]. According to security experts and consultants, Bruce Schneider recognizes enforcement challenges, especially concerning the international supply chain for IoT devices [60]. Schneier recommends establishing a clear legal definition of unsafe devices. These procedures include legal policies on penalties for non-compliance. These policies focus on creating a certification and labeling authority to enhance device safety confidence. They apply to international partners in the European Union. In addition, they help develop and enforce device security standards. Tens of billions of IoT devices are currently deployed worldwide [20]. A significant problem with this ever-growing collection is its lack of physical, software, and development standards. The European Union is ahead of the United States in developing cybersecurity measures for Consumer Internet of Things Baseline Requirements [61].

We recommend these steps be implemented for IoT devices [61]:

- No universal default passwords.
- Implement a means to manage reports of vulnerabilities.
- Keep software updated.
- Securely store sensitive security parameters.
- Communicate securely.
- Minimize exposed attack surfaces.
- Ensure software integrity.
- Ensure that personal data is secure.
- Make systems resilient to outages.
- Examine system telemetry data.
- Make it easy for users to delete user data.
- Make installation and maintenance of devices easy.
- Validate input data.

These concepts, ideas, and theories align with many other efforts to develop recommendations for IoT deployments. However, the problem is not recognizing the need but the enforcement. The chapter discusses ways to build safer, more straightforward, and robust IoT security solutions.

19. Conclusion and recommendations

Introducing ZT to traditional network security architectures such as IoT platforms requires an all-hands-on-deck effort from leaders, security practitioners, and stakeholders. ZT is a term that industry leaders and cybersecurity experts have used to describe a security architecture coupled with design principles and strategic methods for cybersecurity. ZT is based on the traditional models' limited security capabilities. From a holistic viewpoint, ZT embraces different security approaches and flavors. Some of these methods include creating perimeter control through complex identifies. These security and technical modalities emphasize robust authentication and identity verification of the users. When implementing ZT into IoT, organizations must establish a roadmap highlighting trust and security policy and how these practices can be managed. Organizations must include Identity and Access Management when adopting and implementing ZT security. These concepts provide other capabilities, such as assessing and updating data in transit. These methods must be applied through identity lifecycle automation with real-time response to access risks. Cybersecurity threats are evolving and remain the biggest concern for industry and government leaders. Data privacy, policies, and regulations are critical to IoT operations and information flow. Excellent policies and regulations are needed to ensure that IoT objects function within proper safeguards. In addition, the launch of IoT technology has resulted in many more portable peripheral devices being manufactured and sold in the marketplace, as organizations must invest in security and monitoring capabilities to protect their IoT systems. Security vendors must develop advanced applications, i.e., TLS and Secure Socket Shell.

ZT and IoT solutions discussed in this chapter reflect an understanding of the challenging security issues. However, even if these involved security solutions may provide acceptable remedies for new and planned implementations, the problems of enforcement and legacy devices remain a de facto challenge for security vendors, experts, and the industry. We strongly recommend the ZT architecture and security approach to current infrastructure projects. Such an approach will prevent unsecured and unmanageable network-connected devices. In addition, industry leaders and cybersecurity experts are committed to evaluating and redesigning their workforce business model to address growing cybersecurity threats and data breaches and harden their corporate infrastructures. Organizations rely on these design principles to establish and eliminate any implicit trust. In addition, they use these concepts as a baseline for validating digital interaction processes. ZT Reference Architecture was developed alongside the "core zero trust logical component." In addition, the reference architecture includes operation and components. Such processes are developed to ensure autonomous on-premises and cloud-based services operation. ZT migration does not merely focus on replacing traditional and/or modern infrastructure. Instead, the approach is based on the organization's cultural changes. This process generally involves organization assets. Resources can be inventoried and maintained within the organization. For many years risks, threats, and vulnerabilities have impacted organizations' infrastructure. ZT security experts continue researching, developing, and deploying new security capabilities to minimize outside and insider threats, risks, and vulnerabilities disrupting the organization's day-to-day assets and resources. Government, corporations, and stakeholders continue to benefit from ZT technology. Yet, ZT security remains the concern of policymakers, users, and organizations. As a result, decision-makers and cybersecurity experts must develop security processes to address continued threats, risks, and vulnerabilities against the organization's devices and infrastructure. ZT capability requires a Continuous Monitoring process and applicability. CM is critical for ZT's adoption, implementation, and functioning. ZT environment requires a trust level function between the system, data flow, user, and application. The genesis of IoT technology gives devices/objects the capability to share data without requiring continuous signature updates or human interference. However, adopting and implementing IoT technology have advantages and disadvantages. IoT is built to provide advanced interconnect solutions and allow physical and virtual devices and objects to interact autonomously as a global infrastructure.

Commercial ZT and IoT offerings, such as those outlined in this chapter from Microsoft, Amazon, and Google, provide an excellent start for addressing complex IoT security issues. IoT security focuses on the organization and individual consumers' protection of IoT resources and related technology assets. Despite this process, more work must be done in the computer and network security domains. Tiffany McDonald [62] notes, "The writing is on the wall. The era of deploying poorly secured IoT devices with 10-year-old unpatched network services is ending. The risks are too high." Instead, industry and government leaders and cybersecurity experts must invest in security solutions, digital certificates, and authentication capabilities. Recent security studies have proven that PKI's digital/public vital certificates serve as an additional security layer that gives IoT devices/objects the protection they need.

References

[1] B. Schneier, Security and the Internet of Things, 2017. https://www.schneier.com/blog/archives/2017/02/security_and_th.html. (Accessed 19 February 2022).
[2] Microsoft, Evolving Zero Trust: How Real-World Deployments and Attacks Are Shaping the Future of Zero Trust Strategies, 2021. https://query.prod.cms.rt.microsoft.com/cms/api/am/binary/RWJJdT. (Accessed 9 March 2022).
[3] Gartner, Zero Trust Architecture and Solutions, 2022. https://www.gartner.com/teamsiteanalytics/servePDF?g=/imagesrv/media-products/pdf/Qi-An-Xin/Qi-An-Xin-1-1OKONUN2.pdf. (Accessed 23 March 2022).
[4] Hewlett Packard Enterprise, Edge Computing Yields Deeper Insight, faster, 2022. https://www.hpe.com/us/en/insights/reports/intelligent-edge-report-1903.html. (Accessed 1 April 2022).
[5] Gartner, Gartner Insights on how to Lead in a Connected World, 2017. https://www.gartner.com/imagesrv/books/iot/iotEbook_digital.pdf. (Accessed 3 April 2022).
[6] PSA Certified, The History of IoT Security, 2021. https://publications.psacertified.org/the-history-of-iot-security/history-in-the-making/. (Online]. Accessed 20 February 2022).
[7] Zero trust security model, Wikipedia, 2022. https://en.wikipedia.org/wiki/Zero_trust_security_model. (Accessed 9 December 2022).
[8] Authentication, 2022. In *Wikipedia* https://en.wikipedia.org/wiki/Authentication. (Accessed 10 December 2022).
[9] K. Raina, Zero Trust Security Explained: Principles of the Zero Trust Model, 2022. https://www.crowdstrike.com/cybersecurity-101/zero-trust-security/#:~:text=Zero%20Trust%20seeks%20to%20address%20the%20following%20key,endpoint%2C%20workload%2C%20etc.%29%20for%20the%20most%20accurate%20response. (Accessed 19 December 2022).
[10] A. Loten, Akamai bets on 'Zero Trust' approach to security, Wall Street J. (2019) (Retrieved 2022-02-17. Accessed 9 December 2022).
[11] S. Marsh, Formalizing Trust as a Computational Concept, 1994. https://scholar.google.co.uk/citations?view_op=view_citation&hl=en&user=Qz73wh4AAAAJ&citation_for_view=Qz73wh4AAAAJ:u5HHmVD_uO8C. (Accessed 20 December 2022).
[12] Infraon. https://infraon.io/blog/history-of-zero-trust-security/. (Accessed 21 December 2022).
[13] Cyolo, ZTNA vs. ZTA vs. ZTAA. https://cyolo.io/blog/ztna-vs-ztaa-vs-zta-which-one-should-you-choose/. (Accessed 23 December 2022).
[14] I. Dobson, J. Hietala, Jericho Forum Declares "Success" and Sunsets, 2013. Opengroup.org. Retrieved 2018-02-27.
[15] The Jericho Forum's Collaboration-Oriented Architecture Paper. Paper http://www.opengroup.org/jericho/COA_v1.0.pdf. Accessed 22 December 2022.
[16] Joanne Cummings "Security in a world without borders", Network World 27 September 2004 "Face it, you've Already Been de-Perimeterized. The Question Now Is, What Are you Going to Do about it?", 2022 (Accessed 23 December 2022).
[17] Jericho Forum. https://collaboration.opengroup.org/jericho/vision_wp.pdf. (Accessed 24 December 2022).

[18] A. Dhatrak, Cyber Security Threats and Vulnerabilities in IoT, 2020. https://www.irjet.net/archives/V7/i3/IRJET-V7I3582.pdf. (Accessed 22 March 2022).
[19] Changeis. https://changeis.com/strategy/understanding-zero-trust-security/. (Accessed 23 December 2022).
[20] O.S. Rose, S.M. Borchert, S. Connelly, NIST Special Publication 800–207 Zero Trust Architecture, 2020, Available https://nvlpubs.nist.gov/nistpubs/SpecialPublications/NIST.SP.800-207.pdf. (last accessed 26 Mar 2022).
[21] NIST SP 800-207 (Second Draft), Zero Trust Architecture. https://nvlpubs.nist.gov/nistpubs/SpecialPublications/NIST.SP.800-207-draft2.pdf. (Accessed 22 December 2022).
[22] TechBeacon. https://techbeacon.com/security/migrating-zero-trust#:~:text=Enhanced%20identity%20governance%20represents%20the%20simplest%20approach%20to,are%20based%20upon%20user%20identity%20and%20device%20fingerprinting. (Accessed 23 December 2022).
[23] Avertum, Zero Trust SIEM Strategy (Role of SIEM in Zero Trust Environment), 2021. https://www.avertium.com/blog/siem-zero-trust-strategy. (Accessed 21 December 2022).
[24] K. Raina (2022). Zero trust security explained: principles of the zero trust mode. What Is Zero Trust Security? Principles of the Zero Trust Model (crowdstrike.com) Accessed 23 December 20.
[25] GTI https://governmenttechnologyinsider.com/the-five-pillars-of-zero-trust-architecture/. Accessed 31 December 2022.
[26] DoD Zero Trust Strategy, DoDCIO. https://dodcio.defense.gov/Portals/0/Documents/Library/DoD-ZTStrategy.pdf. (Accessed 24 December 2022).
[27] M. Schneider, B. Hippchen, S. Abeck, M. Jacoby, R. Herzog, Enabling IoT Platform Interoperability Using a Systematic Development Approach by Example. https://cm.tm.kit.edu/download/integration_of_iot_platforms.pdf#:~:text=To%20enable%20interoperability%20across%20IoT%20platforms%20multiple%20levels,teroperability.%20Technical%20interoperability%20enables%20the%20exchange%20of%20data. (Accessed 27 March 2022).
[28] I. Cloudflair, Inside the Infamous Mirai IoT Botnet: A Retrospective Analysis, 2017, Available https://blog.cloudflare.com/inside-mirai-the-infamous-iot-botnet-a-retrospective-analysis/. (last accessed 2 Apr 2022).
[29] A. Nordrum, Popular Internet of Things Forecast of 50 Billion Devices by 2020 Is Outdated, 2016. https://spectrum.ieee.org/popular-internet-of-things-forecast-of-50-billion-devices-by-2020-is-outdated. (Accessed 18 February 2022).
[30] Norton, Zero Trust on the Endpoint—Extending the Zero Trust Model from Network to Endpoint with Advanced Endpoint Protection, 2015. https://webobjects.cdw.com/webobjects/media/pdf/paloalto/Zero-Trust-on-the-Endpoint.pdf. (Accessed 3 April 2022).
[31] S. Holger, Endpoint and IoT Zero Trust Security Report, Pulse Security, 2020. https://www.wsta.org/wp-content/uploads/2020/11/PS-2020-Endpoint-IoT-Zero-Trust-Security-Report-14.pdf. (Accessed 1 April 2022).
[32] R. Dsouza, How to Implement Zero Trust IoT Solutions with AWS IoT. [Online], 2021, Available: https://aws.amazon.com/blogs/iot/how-to-implement-zero-trust-iot-solutions-with-aws-iot-3/. (last accessed 14 Mar 2022).
[33] A. Weinert, Traditional Perimeter-Based Network Defense Is Obsolete—Transform to a Zero Trust Model, 2019. https://www.microsoft.com/security/blog/2019/10/23/perimeter-based-network-defense-transform-zero-trust-model/. (Accessed 18 March 2022).
[34] SailPoint, Implementing Zero Trust Security, 2021. https://www.sailpoint.com/solutions/zero-trust/?utm_source=bing&utm_medium=cpc&utm_campaign=*Zero%20Trust

%20AMS&utm_content=82463622871844&utm_term=Improving%20compliance%20in%20Zero%20Trust&utm_id=392798924&msclkid=bb45e31113941be62b8c82e339690b99. Accessed 23 February 2022.
[35] J.M.C.B. Da Silva, Attack and Intrusion Detection on the IoT, Dissertation, Universidade de Coimbra, 2018.
[36] J.H. Koo, Y.G. Kim, Interoperability of device identification in heterogeneous IoT platforms, in: Proceeding of the IEEE International Computer Engineering Conference (ICENCO), Cairo, Egypt, 2017 (Accessed 19 February 2022).
[37] Poneman Institute, Global PKI and IOT Trends Study, 2021. https://www.entrust.com/-/media/documentation/reports/2021-pki-iot-trends-study-executive-summary-re.pdf. (Accessed 24 March 2022).
[38] A. Ometov, S. Bezzateev, N. Makitalo, T. Mikkonen, Y. Kooucheryavy, S. Andreev, Multi-Factor Authentication: A Survey, 2018. https://www.researchgate.net/publication/322288752_Multi-Factor_Authentication_A_Survey. (Accessed 3 April 2022).
[39] Multifactor Authentication (MFA), And Single Sign-on (SSO), 2022. https://support.payapps.com/en/articles/6013443-faqs-multifactor-authentication-mfa-and-single-sign-on-sso#:~:text=MFA%20is%20often%20used%20interchangeably%20with%20two-factor%20authentication,all%20of%20their%20resources%20with%20a%20single%20authentication. Accessed 19 December.
[40] K. Patel, S. Patel, P.G. Scholar, C. Salazar, Internet of Things-IOT: Definition, Characteristics, Architecture, Enabling Technologies Application & Future Challenges, 2016. https://www.researchgate.net/publication/330425585_Internet_of_Things-IOT_Definition_Characteristics_Architecture_Enabling_Technologies_Application_Future_Challenges. (Accessed 8 March 2022).
[41] Intel, Policy Framework for the Internet of Things, 2014. https://www.intel.com/content/dam/www/public/us/en/documents/corporate-information/policy-iot-framework.pdf. Accessed 6 April 2022.
[42] SocketXP, How to Remote Access IoT SSH over the Internet, 2021. https://www.socketxp.com/iot/how-to-remote-access-iot-ssh-over-the-internet/. (Accessed 18 March 2022).
[43] Palo Alto Networks, Unit 42 IoT Threat Report, 2020. https://iotbusinessnews.com/download/white-papers/UNIT42-IoT-Threat-Report.pdf. (Accessed 2 April 2022).
[44] I. Lee, Internet of Things (IoT) Cybersecurity: Literature Review and IoT Cyber Risk Management, School of Computer Sciences, College of Business and Technology, Western Illinois University, Macomb, IL, USA, 2020.
[45] M. Thomas, 13 IOT Security Companies you Should Know, 2019, Available online: https://builtin.com/internet-things/iot-security-companies-startups. (Accessed 12 December 2022).
[46] C. Hsu, J.C. Lin, Exploring factors affecting the adoption of internet of things services, J. Compute. Inf. Syst. 58 (2018) 49–57 (Accessed 11 December 2022).
[47] Markets and Markets, IoT Security Market Worth $35.2 Billion by 2023, 2019, Available online: https://www.marketsandmarkets.com/PressReleases/iot-security.asp. (Accessed 13 December 2022).
[48] DIGI International (N.D.) Strengthening Security in Embedded IoT Solutions. https://www.digi.com/pdf/embedded_security-in-iot_wp.pdf. Accessed 2 April 2022.
[49] AWS, IoT ExpressLink (Preview), 2022. https://aws.amazon.com/iot-expresslink/. (Accessed 27 March 2022).
[50] AWS IoT, 2022. https://aws.amazon.com/iot/. (Accessed 1 April 2022).
[51] Tupid, IoT Comparison: AWS vs. Azure vs. Google vs. IBM, 2019. https://tudip.com/blog-post/comparing-iot-services-with-aws-vs-azure-vs-google-vs-ibm/#:~:text=Pricing-wise%2C%20Azure%20IoT%20is%20cheaper%20than%20the%

20AWS,in%20the%20cloud%20which%20communicate%20with%20each%20other. (Accessed 2 April 2022).
[52] WireFlow (N.D.) How to Connect an IoT-Device to Google Cloud Platform using WireQueue MQTT Toolkit. https://www.wireflow.com/wp-content/uploads/2021/03/AB0005-128-AN20-Google-IoT-How-To.pdf. Accessed 10 February 2022.
[53] W. Opaska, The Default Password Threat, 2005. https://www.giac.org/paper/gsec/317/default-password-threat/100889. (Accessed 3 April 2022).
[54] IANS, Smart Light Bulbs Can Hack your Personal Information, 2019. https://gulfnews.com/technology/smart-light-bulbs-can-hack-your-personal-information-1.1571846229201. (Accessed 22 February 2022).
[55] M. Fagan, J. Marron, K. Brady, B. Cuthill, K. Megas, R. Herold, D. Lemire, B. Hoehm, NIST Special Publication 800–213: IoT Device Cybersecurity Guidance for the Federal Government: Establishing IoT Device Cybersecurity Requirements, 2021. https://nvlpubs.nist.gov/nistpubs/SpecialPublications/NIST.SP.800-213.pdf. (Accessed 25 February 2022).
[56] PSA Certified, History of IoT Security, 2022. https://publications.psacertified.org/the-history-of-iot-security/history-in-the-making/. (Accessed 3 April 2022).
[57] R. Srinivas, IoT Security Incidents that Make you Feel Less Secure, 2020. https://cisomag.eccouncil.org/10-iot-security-incidents-that-make-you-feel-less-secure/. (Accessed 21 March 2022).
[58] A. Fowler, J. Goel, M.M. Hodges, Compliance Planning for California IoT Security Requirements, 2020. https://www.hwglaw.com/compliance-planning-for-california-iot-security-requirements/. (Accessed 19 February 2022).
[59] P. Muncaster, Government Promises IoT Security Enforcement Body, 2020. https://www.infosecurity-magazine.com/news/government-promises-iot-security/. (Accessed 17 March 2022).
[60] H. Kim, T. Herr, B. Schneier, Enforcing Security on the Global IoT Supply Chain, 2020. https://www.atlanticcouncil.org/wp-content/uploads/2020/06/Reverse-Cascade-Report-web.pdf. (Accessed 13 February 2022).
[61] ETSI, Cyber Security for Consumer Internet of Things: Baseline Requirements, 2020. https://www.etsi.org/deliver/etsi_en/303600_303699/303645/02.01.01_60/en_303645v020101p.pdf. (Accessed 11 February 2022).
[62] T. McDonald, IoT Security Compliance and Enforcement | ReFirm Labs. 2021 https://automatemyhome.com/iot-security-compliance-and-enforcement-refirm-labs/ (Accessed 2 April 2022).

About the authors

Dr. Marcus Tanque is a principal researcher, professional, and scholar. Dr. Tanque has worked with government-industry stakeholders. Marcus is an independent researcher, author, editorial review board member, and referee for several books, journal articles, and published chapters. He holds a Ph.D. in Information Technology with a dual Specialization in Information Assurance and Security and an M.S. in Information Systems Engineering. In his spare time, Dr. Tanque enjoys helping and mentoring others. He is also an enthusiastic reader and researcher who collaborates with like-minded professionals.

Harry J. Foxwell is Associate Professor at George Mason University's Department of Information Sciences and Technology. He earned his doctorate in Information Technology in 2003 from George Mason University's Volgenau School of Engineering (Fairfax, VA), and has since taught graduate courses there in big data analytics and ethics, operating systems, computer architecture and security, cloud computing, and electronic commerce.

CHAPTER FOUR

A realtime fingerprint liveness detection method for fingerprint authentication systems

Chengsheng Yuan[a,b], Qianyue Zhang[c], Sheng Wu[d], and Q.M. Jonathan Wu[a,e]

[a]School of Computer Science, Engineering Research Center of Digital Forensics by Ministry of Education, Nanjing University of Information Science and Technology, Nanjing, China
[b]Key Laboratory of Public Security Information Application Based on Big-Data Architecture by Ministry of Public Security, Zhejiang Police College, Hangzhou, China
[c]School of Computer Science, Nanjing University of Information Science and Technology, Nanjing, China
[d]Operating Unit on Policy-Driven Electronic Governance, United Nations University, Guimarães, Portugal
[e]Department of Electrical and Computer Engineering, University of Windsor, Windsor, ON, Canada

Contents

Advances in Computers, Volume 131
ISSN 0065-2458
https://doi.org/10.1016/bs.adcom.2023.04.004

Abstract

Fingerprint authentication is widely used in various intelligent devices. Fraudulent Attack (FA) using forged fingerprints is one of the most seen security threats in fingerprint authentication. Fingerprint Liveness Detection (FLD), a system to detect if a fingerprint is natural (from a live person present at the point of capture) or fake (from a spoof artifact or lifeless body part), is an essential step before fingerprint authentication. This paper proposes a lightweight and real-time FLD method based on a comprehensive learning system with a Uniform Local Binary Pattern (ULBP) for Fingerprint Authentication Systems (FAS). Our approach contains three steps: First, perform the Region of Interest (ROI) extracts for fingerprint samples to eliminate data noise. Second, to construct distinguishable texture features by ULBP descriptors as the Broad Learning System (BLS) input. ULBP reduces the variety of binary patterns of fingerprint features without losing any critical information. Third, the extracted features are fed into the BLS for subsequent training. The BLS is a flat network that transfers and places the original input as a mapped feature in feature nodes, generalizing the structure in augmentation nodes. The results from the experiment with the five public fingerprint datasets (LivDet 2011, 2013, 2015, 2017, and 2019) show that, compared to other advanced models, our method has faster speed, smaller size, and higher performance. Our method can achieve excellent FLD results in non-high-performance equipment, which is critical for its application in mobile devices.

1. Introduction

With the fast development of big data and artificial intelligence, the security problems [1] related to data privacy protection have drawn increased attention. To ensure the information system is not to be used illegally, verifying the authenticity of the user's identity is essential. Traditional authentication methods use usernames and passwords, and their flaws are apparent; that is, the usernames and passwords are easily get lost, stolen, or deciphered [2]. There is a pressing need to develop more secure and convenient methods for identity authentication. Biometric technologies, including fingerprint recognition, gait recognition, face recognition, voiceprint recognition, etc., are increasingly adopted in various settings, such as the financial sector, social security field, and judicial authentication. The characteristics of human biometrics are uniqueness, accessibility, and universality, which can support identity authentication. According to the China Biometric Technology Market Research and Development Trend Analysis Report (2021–27) [3], among the existing biometric applications as of 2014, fingerprint recognition [4], as the most mature and mainstream authentication

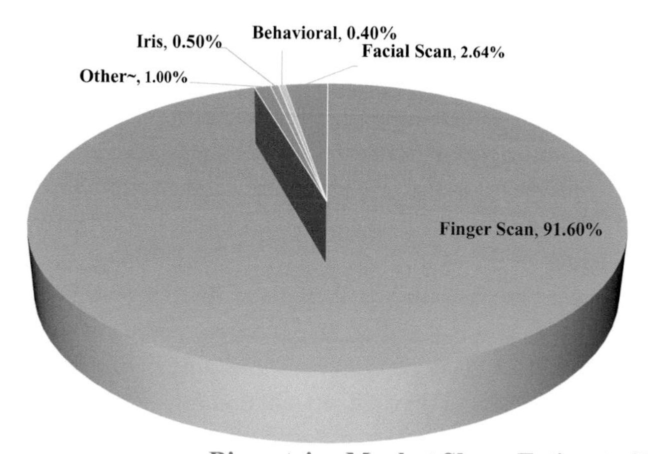

Fig. 1 Biometric application market share estimation. *Source: China Biometric Technology Market Research and Development Trend Analysis Report (2021–27), www.Cir.cn*

technology, accounts for 91.6% of the market share [5–8]. It is widely used in mobile payment, attendance systems, and access control identification, as shown in Fig. 1.

Despite its high potential, fingerprint authentication systems face security risks [9–12], especially while new technologies, such as high-resolution simulation, 3D printing, and generative adversarial technique [13], have emerged in recent years. Fingerprint Authentication Systems (FAS) is vulnerable to active attacks, such as attacks by a synthetic fingerprint generation method through the technologies mentioned above or attacks by using specific materials, such as silica gel, alum, barium alum, plasticine to collect fingerprints [14,15]. To mitigate these risks, Fingerprint Liveness Detection (FLD) [2,16,17] approaches are proposed as the primary countermeasure to protect FAS [18]. Many activities intending to accelerate FLD innovation have been held in recent years, attracting a lot of attention from academia and industries, including the International Fingerprint Liveness Detection Competition (IFLDC) [19–23] and Fingerprint Verification Competition (FVC) [24], etc.

Among many detection algorithms, studies on Deep Learning based FLD [9,25–27] have achieved considerable progress. However, some constraints remain in their application. For example, with limited storage space and computing power, mobile devices cannot operate those complex Neural Network algorithms. Also, the existing Convolution Neural Network

(CNN) models contain many parameters. Thus, the training time is long, and some even take a few months. Broad Learning System (BLS), which is applied to FLD for the first time, can realize shortened model training time; however, its vast network structure remains a challenge for achieving satisfactory real-time performance. We propose a lightweight and real-time FLD method based on broad learning with a Uniform Local Binary Pattern (ULBP), which suits mobile terminals to meet this challenge. Our method contains three steps: First, we use the Region of Interest (ROI) method to extract the effective fingerprint region and remove redundant information. Second, we use the ULBP to extract fingerprint features. Third, the extracted features are input as the BLS [27] for subsequent training. Without multi-layer connections, the BLS does not need to use gradient descent to update network weights; thus, training of this model is much faster than Deep Learning model. We increase the width of the network, so it can adapt to different datasets and improve detection performance. Compared with adding layers in a Deep Learning network, the computational workload of increasing the width is negligible.

To sum up, the main contribution of this paper is to establish a method that is innovative from the following perspectives:

(1) Using ULBP to extract the feature vector to reduce the impact of high-frequency noise. As a result, more discriminative fingerprint features (real or fake) can be retained. The performance of FLD remains the same while the training time has been reduced.

(2) We apply BLS to FLD for the first time. This model does not require GPU training. It has outstanding real-time performance. Therefore, it is more suitable for mobile terminals, which have limited computing power.

(3) BLS does not require retraining the entire system with new data. When the system collects new data, the existing network structure will be updated to reflect the system's integrity. Experimental results show that the performance of this system is better than existing models, and the average training time is about 1 s.

The rest of our paper is as follows. In Section 2, the concepts related to fingerprints and is reported. In Section 3, the classification of FLD methods is introduced. After that, the construction of our FLD system is presented in Section 4, including our model architecture and related theories. In Section 5, we analyze the experimental results. Then we conclude our study and further research needed in the future in Section 6.

2. Concepts related to fingerprint and FIS

Fingerprint Identification Systems [28] are widely used nowadays, such as in access control, attendance records, and digital payment. Some synthetic fingerprint films can compromise these systems. Therefore, developing methods for distinguishing actual fingerprints from artificial ones is critical.

2.1 Fingerprint pattern

The early use of fingerprints can be traced back to the ancient Babylonians and the Chinese. Fingerprints are the ridges of skin on the belly of the end of a human finger [29]. They are formed naturally in the process of human evolution. The shape of fingerprints does not change with ontogeny, except for some changes in invisibility. The fingerprints are "different and constant throughout life." There are three basic shapes of fingerprint patterns: whorl, arch, and loop, as shown in Fig. 2 [30].

2.2 Fingerprint feature

As shown in Fig. 2, to distinguish the two fingerprint images, the key is in the details, including ridgeline, sweat pore, grain shape, details points, lines, etc. We can store the features of these details for fingerprint identification, which can reflect the whole fingerprint image.

To confirm the uniqueness of a fingerprint, two categories of features are needed: general features and local features [28]. These two categories of features are the basis of fingerprint image identification. Available features include emphasis (located at the center of the fingerprint pattern of gradual),

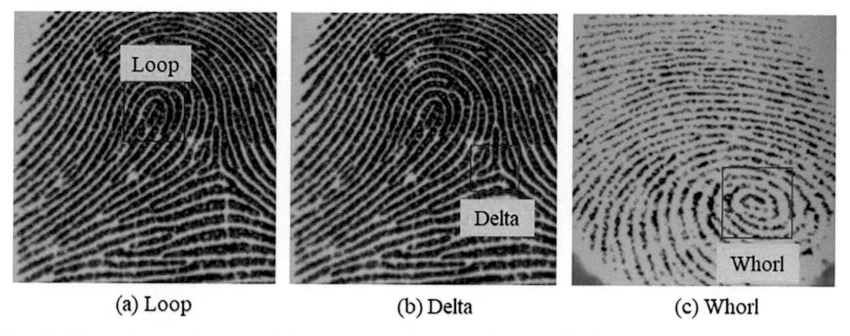

Fig. 2 Three basic shapes of fingerprint pattern. *Source: LivDet dataset.*

triangle points (emphasis is located at the start of the first branch point or a breakpoint, or two lines converging at, isolated points, turning or pointing to these singularities), the number of lines (the number of lines in a fingerprint); Local features are the detailed features of fingerprints. The direction, curvature, and node position at feature points are essential indicators for distinguishing different fingerprints. As shown in Fig. 3, the features of fingerprints can be categorized into three levels: First-level feature, Second-level feature, and Third-level feature [29,30].

First-level feature: As shown in Fig. 4, First-level features are mainly composed of fingerprint ridges, including three basic shapes, whorl, delta, and loop. Fingerprints are made by the different directions of the subcutaneous tissue pressure on the finger epidermis. In recent years, with the introduction of advanced technologies, such as Generative Adversarial Networks (GAN, Conditional GAN, etc.), attackers can simulate complex and rare fingerprints.

Second-level feature: Second-level features refer to the correlation details of ridges and ridgelines. As shown in Fig. 5, the standard detailed features mainly include ridge ending, bifurcation, lake, independent ridge, spur, point or island, etc.

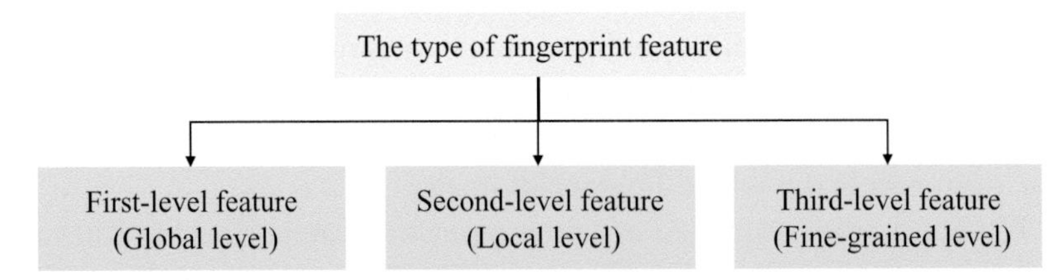

Fig. 3 The type of fingerprint features. *Source: Author.*

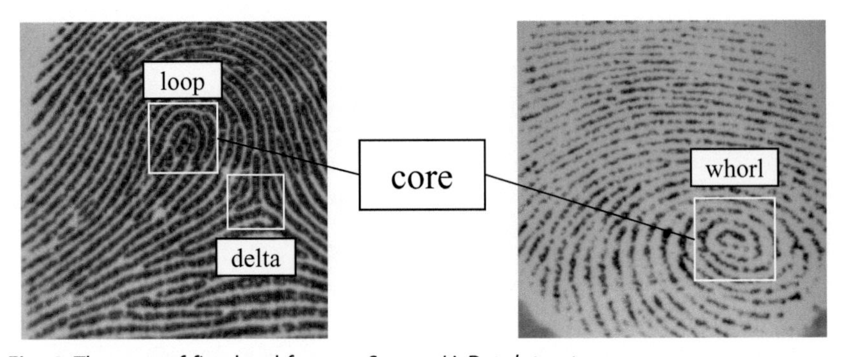

Fig. 4 The type of first-level feature. *Source: LivDet dataset.*

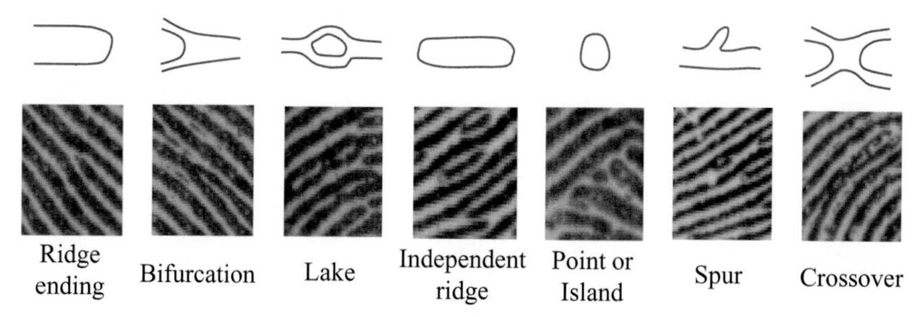

Fig. 5 The type of second-level feature. *Source: LivDet dataset.*

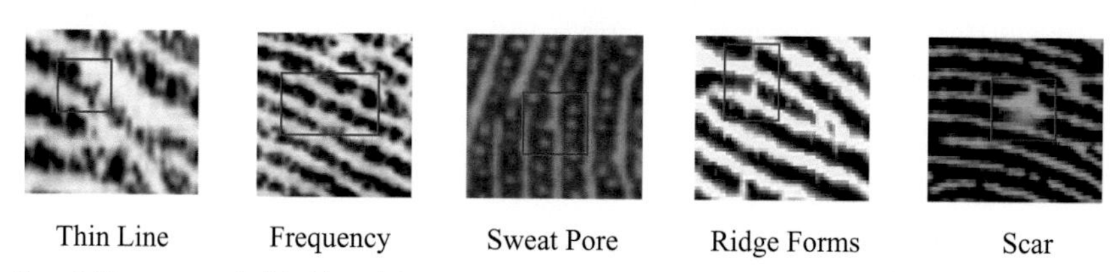

Fig. 6 The type of third-level feature. *Source: LivDet dataset.*

Third-level feature: To distinguish the two levels of features mentioned above by using a computer, the third-level feature is established. In this method, the features of fingerprint images are learned manually or automatically by computer and image processing technology (as shown in Fig. 6) and stored in the feature database. The fingerprint feature template dataset is constructed to facilitate the subsequent feature matching.

2.3 Fingerprint matching

Fingerprint matching uses algorithms to find representative features in fingerprint images and verify the authenticity of users' identities when comparing these features. As shown in Fig. 7, fingerprint matching involves two processes. One is user registration, i.e., the construction of the user feature template set:

First, the user's fingerprint is collected by a sensor or scanner. Due to the influence of hardware or the surroundings, the fingerprint images collected are impure, so pre-processing is performed.

Then, a local or global feature representation algorithm is used to extract fingerprint features, and feature vectors are stored in the feature data set to complete user registration.

The other process is user verification:

First, its operation before matching is similar to the registration process. Then, the feature vector passes through the matcher, the feature vector

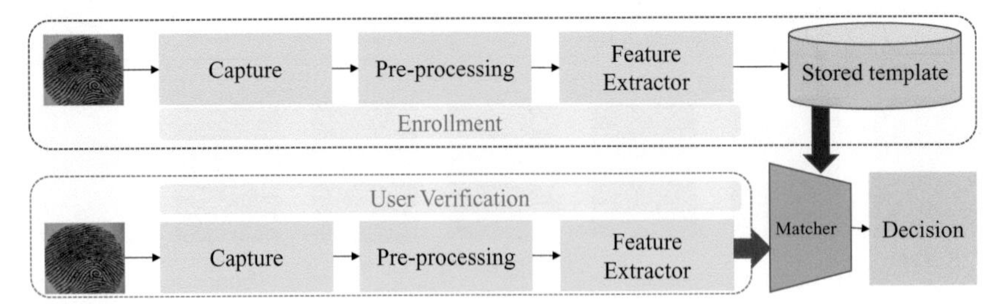

Fig. 7 The general process of fingerprint matching. *Source: Author.*

between the fingerprint to be verified, and the template fingerprint will be matched. If a similar feature vector is identified, the match is successful.

2.4 Fingerprint recognition technologies

Fingerprint Authentication Technologies (FAT) [31] apply intelligent algorithms to find representative features of fingerprint images and verify the authenticity of the user's identity by comparing these unique features. Fingerprint authentication usually classifies the overall features of the fingerprint, such as triangular points and patterns, and then uses local features, such as direction and location, to identify the user. Therefore, in fingerprint identification [30], feature points (Minutiae) should be extracted from the collected images first, and then a fingerprint feature database should be constructed. Finally, the image features to be measured are extracted and matched in the feature data set. The detailed realization process of fingerprint identification is shown in Fig. 7.

Nowadays, the development of big data and artificial intelligence enables the fast implementation of intelligent services, including intelligent medical treatment, virtual personal assistance, etc. Meanwhile, the need for strengthening data security is becoming more and more pressing [32–34]. FAT is drawing more attention while the need increases.

3. FLD and classification of FLD methods

3.1 Definition of FLD

Fingerprint Liveness Detection (FLD) (or Spoof Fingerprint Detection [35]), the approach to identify whether fingerprint images are real or fake, provides new technical support for combating fake fingerprint spooking attacks. Unlike in other image classification tasks, the patterns of authentic and counterfeit fingerprints are very similar. The differences are mainly in

Fig. 8 Real and fake fingerprint samples. Fake fingerprints cannot be identified with the naked eye. *Source: LivDet dataset.*

subtle texture structures that the differences are invisible to the naked eye, as shown in Fig. 8. So far, most fake fingerprint detection methods based on texture draw on the feature extraction of natural images without exploring the essential difference between real and fake fingerprints [31]. These methods are of constraints: for example, while the direction of the fingertip pressing on the scanner is random, and the collected fingerprint is easily disturbed by strong light, the general texture feature algorithm does not have rotation or illumination invariance. For another example, there are many kinds of artificial fingerprint materials (such as capacitive glue, milky white glue, silica gel, etc.). The existing methods do not achieve high detection accuracy when the types of materials are unknown. To overcome these constraints, this paper proposes an effective approach to improve the robustness of FAS and enhance its applicability.

The process of FLD consists of two stages: Training and Testing, as shown in Fig. 9. First, the image is input to the core algorithm; then, the liveness of the fingerprint authenticated is detected by using a black box test; finally, the results are generated. If the fingerprint is from an alive person, it will be sent to the next authentication phase.

3.2 Classification of FLD methods

Research on FLD can be grouped into two types: hardware-based and software-based.

The hardware-based FLD requires a professional sensor at the input end of the fingerprint scanner to detect physiological traits of the human body, such as blood oxygen content, multispectral images, finger veins, etc. [36]. Fingerprint liveness is directly detected by subdermal ridges with high-resolution sensors [37], such as Lumidigm multispectral or Compact Imaging multireference optical coherence tomography. The cost of

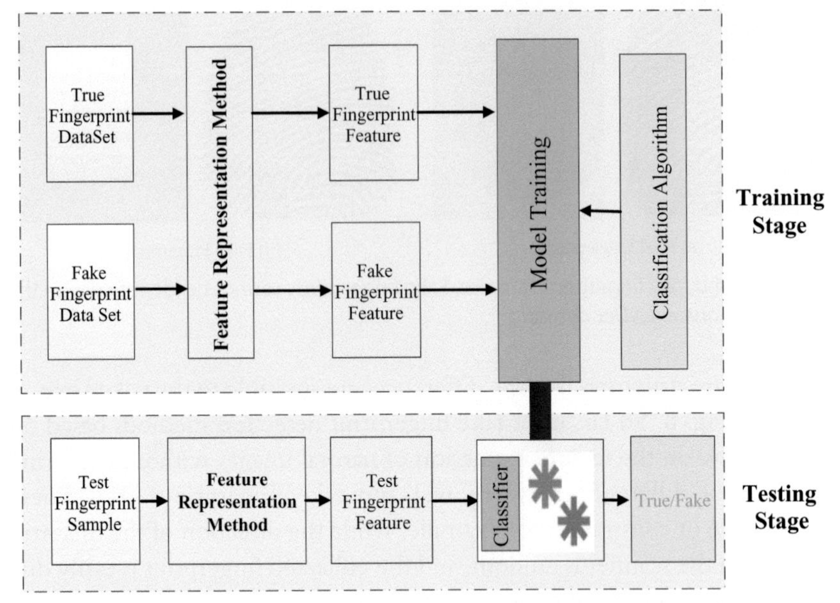

Fig. 9 General process of fingerprint liveness detection. *Source: C. Yuan, Q. Cui, X. Sun, Q.M.J. Wu, S. Wu, Fingerprint liveness detection using an improved CNN with the spatial pyramid pooling structure, Adv. Comput. Elsevier, 120 (2021) 157–193.*

hardware-based FLD is high. It also requires the users to cooperate in the operation process.

The software-based FLD has no such constraints [38,39]. The software is relatively convenient to update and maintain in the later stage. Now it is the mainstream research direction in FLD. Software-based FLD can be divided into three categories: heuristic, texture feature, and Deep Learning.

3.2.1 FLD based on Heuristic

The Heuristic (FLD) refers to recognizing authentic and fake fingerprints with the help of an alive body. Heuristic FLD techniques are divided into skin-deformation-based FLD methods, perspiration-based FLD methods, pore-based FLD methods, and image quality-based FLD methods.

Many sweat pores with relatively stable shape, location, size, and density are distributed on the mastoid stripes of authentic fingerprint images. They are the source of sweat production. The active state of perspiration pores is a unique sign of a natural fingerprint. A number of methods have been proposed based on this condition. [2,40–44]. Although the heuristic FLD method research has clear principles and theoretical basis, in order to observe and analyze the fingerprint activity, it is necessary to analyze the gray variation produced by two or more images. At present, the accuracy of heuristic detection methods needs to be further improved.

3.2.2 FLD based on texture feature

Authentic and fake fingerprint images have different texture patterns [45], and the subtle differences between the two can be used as an important clue for detection. Since the main component of the fake fingerprint material is an organic polymer, it is easy to agglomerate into blocks during the processing process, leaving the fake fingerprint's surface not as smooth as the natural one's. To check the authenticity of the fingerprint, Moon [46] used wavelet transform to analyze the surface roughness of authentic and fake fingerprint images, reconstructed the original fingerprint image by two wavelet transforms and soft threshold processing for six details, and calculated the noise residuals of two fingerprint images before and after denoising. Abhyankar et al. [47] detected forged fingerprints by analyzing the frequency of cross ridges in fingerprint images. The advantage of this methods is that only one single fingerprint image is needed. Nikam [48] proposed a forged fingerprint detection algorithm, which combines Local Binary Pattern (LBP) and wavelet transform to improve the detection performance. LBP was used to describe the fingerprint image's local texture features. The wavelet transform calculates the fingerprint ridge's direction field and frequency information. Tan [49], Gragnaniello [50], Gragnaniello [51], Dubey [52] and Gottschlich [53] also developed their texture feature extraction methods of FLD, and achieved satisfactory detection performance compared with other research algorithms.

3.2.3 FLD based on deep learning

Most of the above FLD methods use feature engineering and rely heavily on experts' knowledge and experience. To improve the precision of FLD, extracting discriminant features is the key, as revealed in Fig. 9. In the recent decades, with the fast development of big data analytics and high-performance computing, Deep Learning has been increasingly used in FLD.

Convolutional Neural Networks (CNNs) is a mainstream Deep Learning architecture [54]. CNNs use relatively little pre-processing compared to other image classification algorithms. The network learns to optimize the filters through automated learning. Nogueira et al. [55] of New York University applied the CNN model to forged fingerprint detection for the first time. They pointed out in their research that due to the network's complex structure and multiple parameters, the over-fitting problem can easily occur when the number of fingerprint samples is limited. Thus, it is necessary to conduct pre-training and parameter fine-tuning for the model. Park et al. [56] first performed two operations of segmentation and sample expansion for fingerprint images and realized true and false identification of fingerprint images by statistical classification results of sample blocks and voting method. Park [25] designed a GRAM module based on the fine texture structure, the critical information for FLD, as he identified. Jung et al. [57] proposed an improved CNN framework in which the mean square regression error was applied to the receptive field to replace the entire connection layer of the network. Wang [58] proposed a forged fingerprint detection model with the voting method. Zhang [59] proposed a lightweight CNN model by improving the residual block of the model. Yuan et al. [12] developed a fingerprint activity detection algorithm with an adaptive residual Neural Network. With an adaptive learning module in the Neural Network, the model's parameters can avoid falling into local optimization. With a texture enhancement scheme, the model's generalization ability can be further improved. Yuan's team [60] also proposed an algorithm based on a semi-supervised self-coding deep network model: First, the parameters of unlabeled samples were pre-trained, then the labeled fingerprint samples were used to fine-tune the parameters and to conduct the final detection. Later, Yuan et al. [61] proposed an algorithm based on the image scale equalization CNN model. This algorithm resolved the constraints on the input image scale of CNN models.

Deep learning models have achieved relatively good results in FLD, but these models still have limitations. For example, complex deep network structure often causes over-fitting problems due to limited manual annotation ability, a small training sample set, and too many parameters. For another example, the detection accuracy of these models will significantly drop when encountering fake fingerprints by unknown materials.

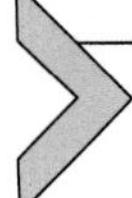

4. Construction of a new LBP FLD

4.1 Uniform local binary pattern

Local Binary Pattern (LBP) [35] descriptor takes 3×3-pixel blocks as the basic unit (or Window). The difference between the central pixel and its adjacent 8 pixels is extracted as local texture feature representation. When the pixel value of the adjacent pixel is less than the pixel value of the central

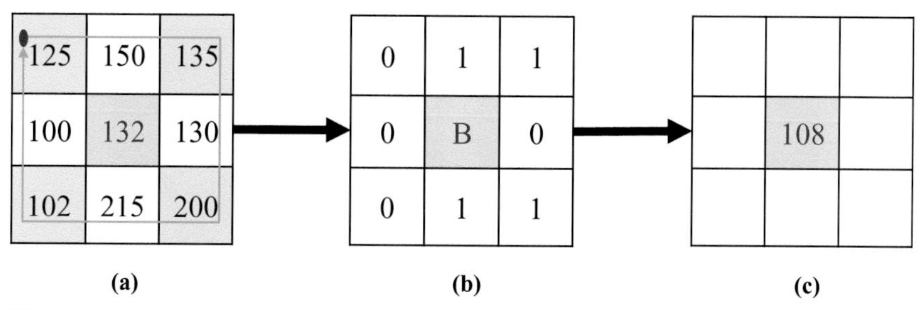

Fig. 10 An example of original local binary pattern. *Source: Author.*

pixel, the value at the position of the adjacent pixel is set as 0. Otherwise, it is set as 1. The pixel point in the upper left c binary form clockwise to obtain the last LBP eigenvalue, as shown in Fig. 10.

$$\begin{bmatrix} x_0 & x_1 & x_2 \\ x_7 & T & x_3 \\ x_6 & x_5 & x_4 \end{bmatrix} \tag{1}$$

The calculation formula of the LBP descriptor is as follows: the calculation rule is set as 0 if the adjacent value is less than the center value T and is set as 1 otherwise. We build an 8-bit binary bitstream to generate decimal numbers as the final LBP feature descriptor.

$$x_i = \begin{cases} 1, x_i \geq T \\ 0, x_i < T \end{cases} \tag{2}$$

$$T = LBP = \sum_{i=0}^{8} 2^i x_i \tag{3}$$

According to the corresponding value at the position of the original pixel block, the LBP feature value is formed into a new image-LBP feature map, as shown in Fig. 11B.

In addition, the adjacent pixels in the LBP operator are not only the surrounding 8 but can be adjusted according to the pixel block. The feature category of the given pixel block (texture features) is 2p, in which p is the number of adjacent pixels. Most LBP feature maps in the given pixel block contain two jumps, including bright spots, dark spots, flat areas, and changed edges, but these transitions already contain the most discriminant information. When the binary number corresponding to LBP has at most two transitions from 0 to 1 or from 1 to 0, the number of the transitions

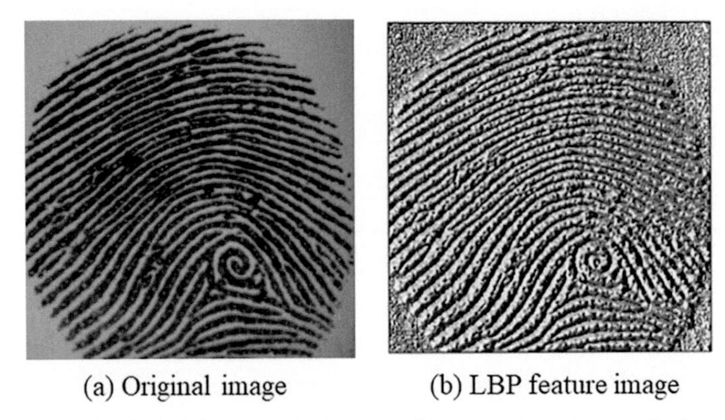

Fig.11 A sample of original image and its LBP feature image. *Source: Author.*

is called an equivalence local binary pattern. For example, 00000000 (0 transition), 00000111 (one transition from 0 to 1), and 10001111 (two transitions) are ULBP. Patterns other than the equivalence pattern are classified into another, called the mixed pattern class. ULBP can be described as:

$$ULBP = \begin{cases} \sum_{p=0}^{p-1} s\left(g_p - g_c\right), ifU(ULBP) \\ p(p-1) + 2, otherwise \end{cases} \tag{4}$$

where *gc* is the pixel of the center, and *gp* is the pixel of the sample in the neighborhood. p is the number of adjacent pixels. s() is signal function, which value is 0 or 1. U() is the number of transitions.

The category of ULBP is $p(p-1)+2$. Thus, the total dimension of texture cpatterns is shortened from $2p$ to $p(p-1)+2$. The LBP feature descriptor does not depend on the image's brightness, and it does not have rotation invariant. If the given pixel block is rotated, the LBP features will change dramatically. The gray value of the circular neighborhood *xi* moves around the circle centered on *T*, as shown in Fig. 12. If *s*(*gp-gc*) are not all 0 or 1, the rotated image will get different LBP features.

Thus, the Rotation-invariant Local Binary Pattern (RLBP) takes an equivalent minimum value after rotation as the final feature representation, and the formula is as follows:

$$RLBP = \min(ROR(LBP, i)), i = 0, 1, \cdots, P-1 \tag{5}$$

where *ROR*(*LBP*, *i*) denotes he equivalent LBP value for all rotations.

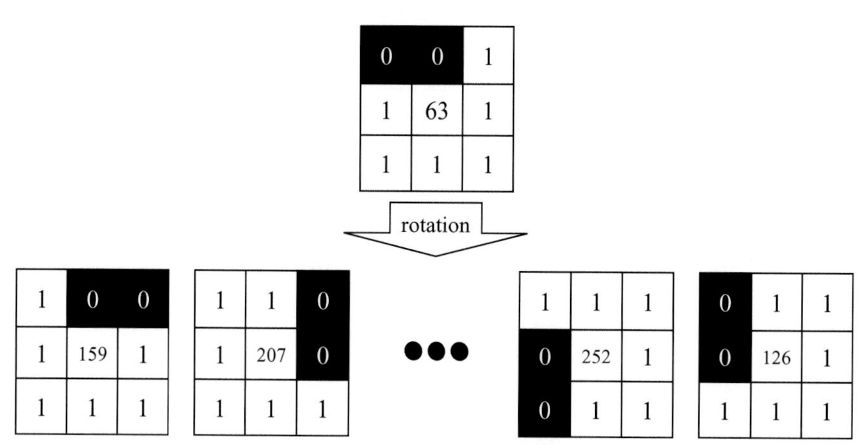

Fig. 12 Equivalent LBP features descriptors after rotation operation. *Source: Author.*

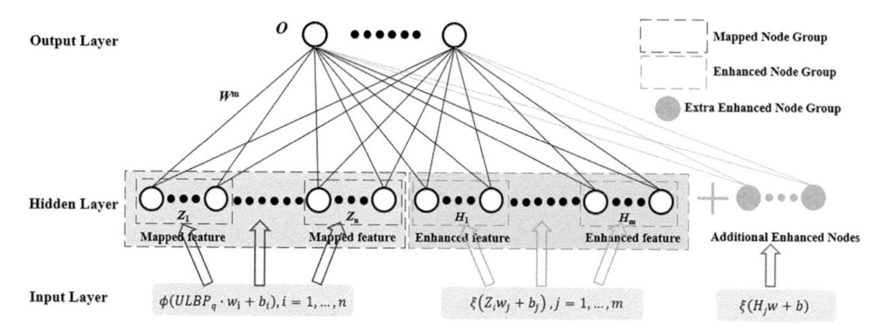

Fig. 13 Our BLS structure with ULBP feature. *Source: Author.*

4.2 Broad learning system

Broad Learning System (BLS) adopts the features mapped as input to the Neural Network nodes [27]. These mapped features are augmented into augmented nodes with randomly generated weights. All mapped features and enhancement nodes are directly connected to the output, and the corresponding output coefficients can be derived by pseudo-inverse. The key to BLS is incremental learning, including augmentation node increment, feature node increment, and input data increment. If the network structure needs to be expanded, there is no need to retrain the entire network. The structure of BLS is shown in Fig. 13.

When passing through the input layer to the hidden layer in this paper, the ULBP will generate some mapped features, and the calculation formula of feature groups is as follows:

$$Z_i = f_i\left(\phi\left(ULBP^* W_i + b_i\right)\right), i = 1, \cdots, n \tag{6}$$

where Zi denotes mapped features and ULBP is the input data of BLS. Wi and bi denote the random weights and bias values, respectively, and n is the number of mapped feature groups. f represents the active function, and fi is the normalization operation to eliminate outliers. Let Z as the concatenation of all the Zi, Z=[Z1, Z2, ..., Zi].

After passing the feature enhancement module, all the mapped features will be further enhanced, and the calculation formula is:

$$H_j = f_i\left(\xi\left(ZW_j + \beta_j\right)\right), j = 1, \cdots, m \tag{7}$$

where Hj denotes enhanced features when the node is j. Wj and bj are the random weights and bias values, respectively. Let H be the concatenation of all the enhancement features Hj, H=[H1, H2, ..., Hm]. After the above operations, the mapped and enhanced features are all input into the output layer. The overall model system is represented by the following formula:

$$O = [Z|H]W^m \tag{8}$$

where O denotes the output of BLS. W denotes the connection weights and can be computed via thec ridge regression: W=(AT′A+c′In+m)−1′AT′O. A is the concatenation operation of the above mapped and enhanced features, and c is a constant parameter. If the model fails to achieve the desired performance, additional boost nodes will be activated to get better performance. Finally, the new weights [27] are:

$$W^{m+1} = \begin{bmatrix} W^m - DB^T O \\ B^T M \end{bmatrix} \tag{9}$$

where

$$D = (A^m)^+ \xi\left(Z^n W_j + \beta_j\right),$$

$$B^T = \begin{cases} (C^+) & if\ C \neq 0 \\ (1 + D^T D)^{-1} B^T (A^m)^+ & if\ C = 0 \end{cases} \tag{10}$$

$$C = \xi\left(Z^n W_j + \beta_j\right) - A^m D; j = 1, \cdots, m \tag{11}$$

Incremental learning only needs to compute the pseudo-inverse of the additional augmented nodes instead of computing the entire (Am+1). Thus, the model training time is shorter, and the number of model parameters is smaller.

5. Experiment

To verify the performance of our method, we first report a brief description of the five public fingerprint databases (LivDet 2011, 2013, 2015, 2017, and 2019) and the operating environment configuration in this section. Then, we introduce the performance evaluation criteria. Lastly, numerical experiments on five well-known public data sets are performed to prove the effectiveness of our algorithm. The research on FLD mainly contains two processes, i.e., the training and testing processes. Details of these processes are shown in Fig. 9. The task of the training process is to build the training model of real and fake fingerprints by learning the different features of real and fake fingerprints. The testing process is also needed to further verify the performance of the various methods, that is, the classification performance of the trained model.

5.1 Computer configuration and fingerprint datasets

Our experimental equipment and operating environment are as the following: Intel(R) Core (TM) i7-10700 CPU @ 2.90GHz, GTX 2070 (8Gb), and 16GB RAM. The operating system is Windows 10 Professional, 64 bits. The model training is conducted on the Python3.7 platform. The detailed computer configuration information is listed in Table 1.

Since the 2000s, many research efforts have been made on FLD. Collecting many fingerprint images is the key to coping with a spoofing attack. The different datasets of images may generate different results, even for the same method. When comparing different FLD methods, it is crucial to set the base on the same fingerprint datasets. University of Cagliari, Italy, and Clarkson University, U.S., jointly launched the first edition of the Fingerprint Liveness Detection Competition 2009 in the context of the 15th International Conference on Image Analysis and Processing (ICIAP 2009). This competition aims to provide a standard testing protocol and a common set of authentic and fake fingerprint data to help competitors

Table 1 Experimental environment.

Hardware condition	Software condition
CPU: Intel@ Core i7-10700 CPU 2.9GHz	OS Version: Windows 10
Memory: 64GB	Run Environment: Python 3.7
Graphics: NVIDIA@ GeForce GTX 2070 (8Gb)	

Source: Author.

compare their respective algorithms. Fingerprint datasets have been released by this competition every 2 years.

To evaluate the performance of our method, we carried out verification with five public fingerprint datasets by the Fingerprint Liveness Detection Competition, i.e., LivDet 2011, 2013, 2015, 2017, and 2019. Details about the five fingerprint datasets are described in Table 2.

Table 2 Fingerprint distribution of the five public LivDet datasets.

Dataset	Sensor (Abbr.)	Train(R/F)	Test(R/F)	Forge fingerprint material type
LivDet 2011	Biometrika(Bio)	1000/1000	1000/1000	EcoFlex,Gelatin,Latex, Silgum,WoodGlue
	DigitalPersona(Dig)	1004/1000	1000/1000	Gelatin,Latex,Playdoh, Silicone,WoodGlue
	Italdata(Ita)	1000/1000	1000/1000	EcoFlex,Gelatin,Latex, Silgum,WoodGlue
	Sagcm(Sag)	1008/1008	1000/1036	Gelatin,Latex,Playdoh, Silicone,WoodGlue
LivDet 2013	Biometrika(Bio)	1000/1000	1000/1000	EcoFlex,Gelatin,Latex, Silgum,WoodGlue
	CrossMatch(Cro)	1250/1000	1250/1000	BodyDouble,Latex, Playdoh,WoodGlue
	Italdata(Ita)	1000/1000	1000/1000	EcoFlex,Gelatin,Latex, Silgum,WoodGlue
	Swipe(Swi)	1221/979	1153/1000	BodyDouble,Latex, Playdoh,WoodGlue
LivDet 2015	Biometrika(Bio)	1000/1000	1000/1500	Ecoflex,Gelatin,Latex, WoodGlue,Liquid, EcoFlex,RTV
	GreenBit(Gre)	996/1000	995/1500	Ecoflex,Gelatin,Latex, WoodGlue,Liquid, EcoFlex,RTV
	DigitalPersona(Dig)	1000/1000	1000/1500	Ecoflex,Gelatin,Latex, WoodGlue,Liquid, EcoFlex,RTV
	CrossMatch(Cro)	1510/1473	1500/1448	BodyDouble,EcoFlex, Playdoh,OOMOO, Gelatin

Table 2 Fingerprint distribution of the five public LivDet datasets.—cont'd

Dataset	Sensor (Abbr.)	Train(R/F)	Test(R/F)	Forge fingerprint material type
LivDet 2017	DigitalPersona(Dig)	999/1199	1700/2028	BodyDouble,Gelatin, Latex,EcoFlex, WoodGlue, LiquidEcoFlex
	GreenBit(Gre)	1000/1200	1700/2018	BodyDouble,Gelatin, Latex,EcoFlex, WoodGlue, LiquidEcoFlex
	Orcathus(Orc)	1000/1200	1700/2040	BodyDouble,Gelatin, Latex,EcoFlex, WoodGlue, LiquidEcoFlex
LivDet 2019	DigitalPersona(Dig)	1000/1000	1020/1224	Gelatin,Latex,EcoFlex, WoodGlue,Mix, LiquidEcoFlex
	GreenBit(Gre)	1000/1200	990/1088	BodyDouble,EcoFlex, WoodGlue,Mix, LiquidEcoFlex
	Orcathus(Orc)	1000/1200	1019/1224	BodyDouble,EcoFlex, WoodGlue,Mix, LiquidEcoFlex

Source: LivDet2011, LivDet2013, LivDet2015, LivDet2017 and LivDet2019.

LivDet 2011 fingerprint database contains 16,056 real and fake fingerprints using four different sensors: Biometrika, DigitalPersona, Italdata, and Sagem. Two types of fingerprints are included in LivDet 2011: training data set with 8020 images and 8036 images. Fig. 14 lists some fingerprint samples from LivDet 2011.

LivDet 2013 database contains 16,853 authentic and fake fingerprints using four different sensors: Biometrika, DigitalPersona, Italdata, and Swipe. Two types of fingerprints are included in LivDet 2013: the training set with 8450 authentic and fake images and 8403 authentic and fake fingerprint images in the testing set. Fig. 15 lists some fingerprint samples from LivDet 2013.

LivDet 2015 database contains 18,922 real and fake fingerprints using four different sensors: GreenBit, DigitalPersona, Italdata, and Biometrika. Two types of fingerprints are included in LivDet 2015: the training set with

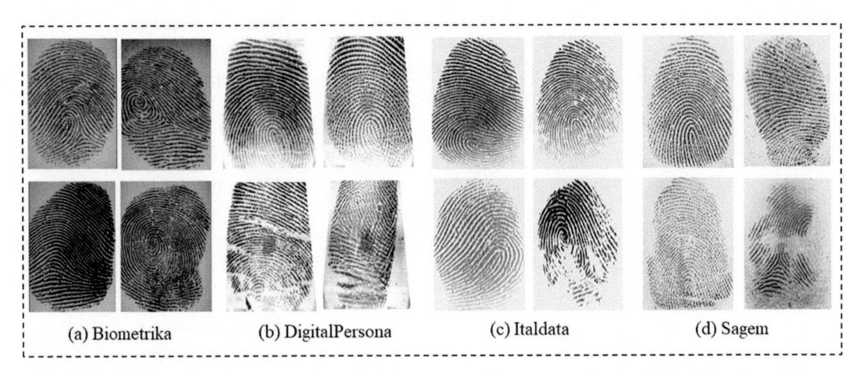

Fig. 14 Fingerprint samples from live (above) and fake (below). Fingerprints acquired with four different sensors: (A) Biometrika. (B) DigitalPersona. (C) Italdata. (D) Sagem. *Source: LivDet 2011 fingerprint database.*

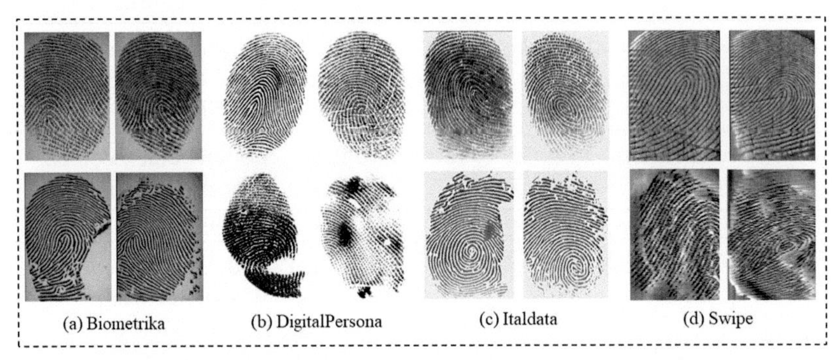

Fig. 15 Fingerprint samples from live (above) and fake (below). Fingerprints acquired with four different sensors: (A) Biometrika. (B) DigitalPersona. (C) Italdata. (D) Swipe. *Source: LivDet 2013 fingerprint database.*

8979 authentic and fake images and 9943 true and fake fingerprint images in the testing set. Fig. 16 lists some fingerprint samples from LivDet 2015.

LivDet 2017 fingerprint contains 17,784 real and fake fingerprints using three sensors: GreenBit, DigitalPersona, and Orcanthus. Two types of fingerprints are included in LivDet 2011: the training set with a total of 6598 images and 11,186 images in the Testing Set. Fig. 17 lists some fingerprint samples from LivDet 2017.

5.2 Performance evaluation criterion

To measure the performance of different FLD methods, a standard performance evaluation criterion is needed. The 2009 fingerprint liveness detection competition [19] defined Average Classification Error (ACE) as the

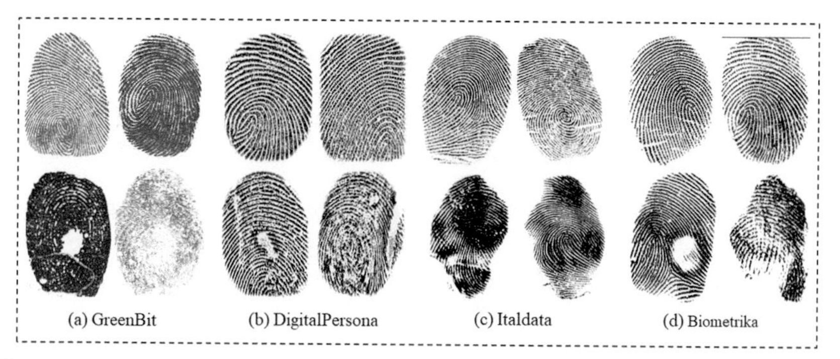

Fig. 16 Fingerprint samples from live (above) and fake (below). Fingerprints acquired with four different sensors: (A) GreenBit. (B) DigitalPersona. (C) Italdata. (D) Biometrika. *Source: LivDet 2015 fingerprint database.*

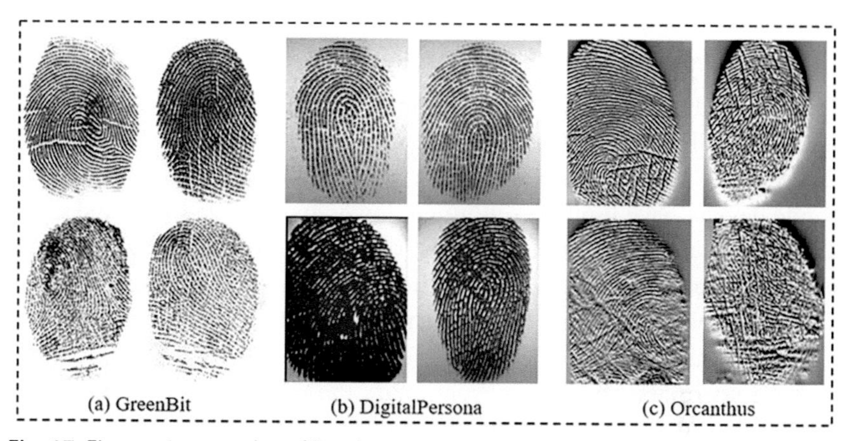

Fig. 17 Fingerprint samples of live fingerprints (above) and fake (below). Fingerprints acquired from three different sensors: GreenBit, DigitalPersona and Orcanthus. *Source: LivDet 2017 fingerprint database.*

standard performance evaluation criterion. In this paper, we use ACE [19] to evaluate and compare the different performances of detection methods. The ACE calculation formula is as follows:

$$ACE = \frac{FAR + FRR}{2} \tag{12}$$

False Accept Rate (FAR) is the probability that a natural fingerprint is wrongly identified as a fake fingerprint, and False Reject Rate(FRR) is the probability of a fake fingerprint being misplaced as a natural fingerprint. The FAR and FRR [62] are calculated as follows:

$$FAR = \frac{misclassified \text{ real fingerprint}}{total \text{ number of real fingerprint}} * 100 \tag{13}$$

$$FRR = \frac{misclassified \text{ fake fingerprint}}{total \text{ number of fake fingerprint}} * 100 \tag{14}$$

According to the formula (15), the ACE range is 0–100. The smaller the ACE is, the better the performance is. To make it easier to view the algorithm's performance, we also adopted the Average Classification Accuracy (ACA) [62] in our previous cwork to measure the correct classification performance of different FLD methods. The calculation formula of ACA is obtained according to the ACE, whose calculation formula is:

$$ACA = 1 - ACE \tag{15}$$

5.3 Experimental results and analysis

The existing Deep Neural Network models are complex, involving many parameters. They demand training time of many hours or even many days. The extensive network structure leads to poor real-time performance. Since mobile devices have limited computing power and storage space, and existing classical and complex deep Neural Networks cannot be deployed and applied to mobile terminals. Our method proposed in this paper can solve this problem well. To verify the performance of our approach, we carry out experiments as follows:

ACE comparison: In our experiment, the relevant parameters of our method are: 30 nodes and 30 windows of the feature mapping layer, 300 nodes and 30 windows of enhance layer, and the number of incremental learning steps is set to 10.

We compared our algorithm with three mainstream Neural Network models, including VGG16 [63], AlexNet [64], and ResNet18 [65], as shown in Tables 3 and 4. The classification results of the above models are shown in these two tables. The difference between Tables 3 and 4 is that the experiments in Table 3 are data expanded, while the fingerprint database in Table 4 is not.

These two tables show that our method obtained the best ACE on the other data sets for data expansion detection results, except for the relatively weak results on data sets Sag, Swi, and Dig. For the extended dataset, shown in Table 4, our algorithm achieves the best results on most datasets.

Average training time comparison: Compared with the three classical CNN models, our method also has advantages in the training time and the number of parameters. As shown in Figs. 18–20, the average training time of our method is the lowest compared with those of VGG16 [63],

Table 3 ACE results on LivDet 2011 and 2013 fingerprint datasets.

		Average classification error ACE in (%)			
Dataset	Model	Bio	Dig	Ita	Sag
LivDet 2011	Ours	1.8	15.6	21.3	16
	VGG16	9.8	13.8	26.2	13.4
	AlexNet	11	13.7	26.6	12.2
	ResNet18	9.4	13	21	15.9
LivDet 2013	Model	Bio	Ita	Swi	
	Ours	2.2	1.3	14.9	
	VGG16	10.6	20.6	12.4	
	AlexNet	14.1	26.6	13.9	
	ResNet18	13.4	13	11.1	

Source: Author.

Table 4 ACE results on LivDet 2015, 2017 and 2019 fingerprint datasets.

		Average classification error in (%)			
Dataset	**Model**	**Cro**	**Dig**	**Gre**	**Hi**
LivDet 2015	Ours	15.2	20.6	11	15.1
	VGG16	17.7	16.4	16.6	33.6
	AlexNet	14.3	33.9	17.7	25.3
	ResNet18	12.6	21.4	11.5	29.4
LivDet 2017		Orc	Dig	Gre	
	Ours	12.2	16	13.4	
	VGG16	28	22.1	34.5	
	AlexNet	24.1	30.6	27.7	
	ResNet18	17.9	24.6	32.4	
LivDet 2019		Orc	Dig	Gre	
	Ours	8.7	28	8.6	
	VGG16	27.7	27.9	28.9	
	AlexNet	22.8	31.6	25.1	
	ResNet18	20.8	38.1	30.7	

Source: Author.

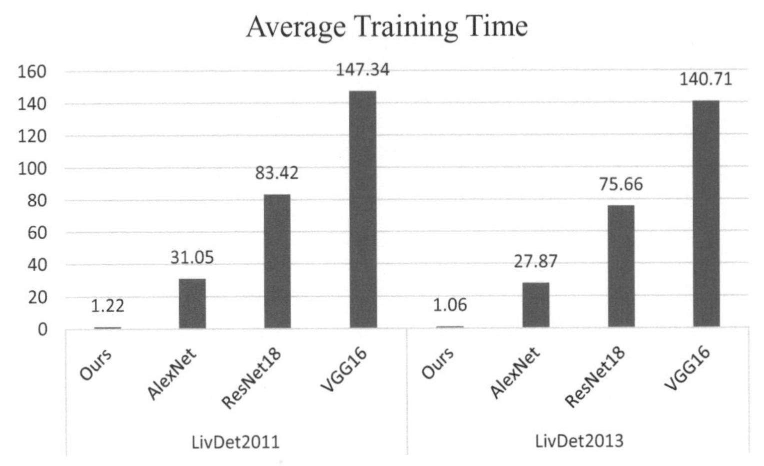

Fig. 18 Average training time in LivDet 2011 and 2013. *Source: Author.*

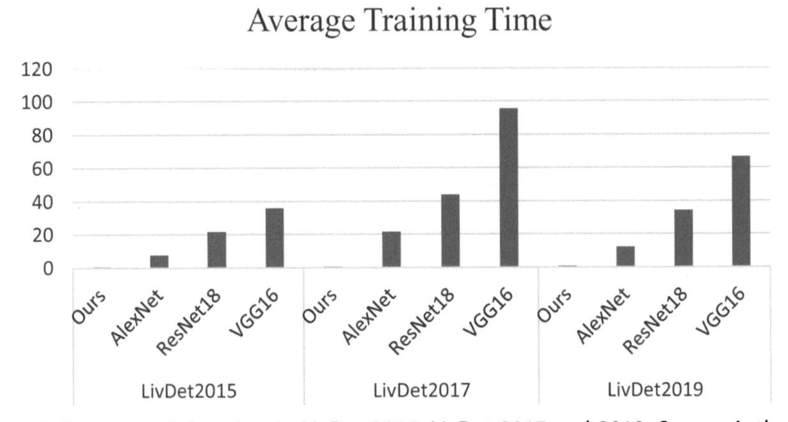

Fig. 19 Average training time in LivDet 2015, LivDet 2017 and 2019. *Source: Author.*

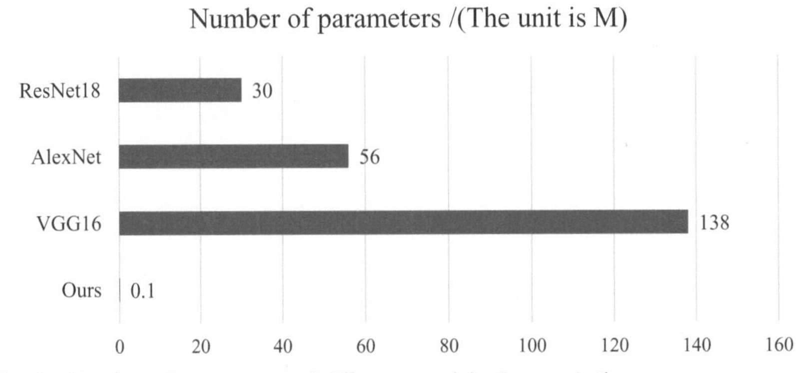

Fig. 20 Number of parameters of different models. *Source: Author.*

AlexNet [64], and ResNet18 [65]. Figs. 18 and 19, together with Tables 3 and 4, demonstrate that the training time of our algorithm is shorter while the error rate remains close.

The number of parameters comparison: Deep Neural Network models require sufficient structural depth to perform. We counted parameters corresponding to different models, as shown in Fig. 20. Our method achieves effective detection results with much fewer parameters, about 300 times less than the second-best. Therefore, our approach requires the least training time and parameters, making it easy to deploy, especially on mobile terminals.

In our algorithm, the key to realizing a lower average training time and fewer model parameters is to choose BLS input as a better feature descriptor. We compared the ACE of three different feature descriptors, as shown in Fig. 21. In general, ACE results of LBP are the best, RLBP results are the weakest, and ULBP results are in between. Considering that the feature dimension of LBP is 256, ULBP is 59, and RLBP is 36, ULBP is selected as the texture feature descriptor in this study. The above experiments also verified that choosing ULBP as the feature descriptor can achieve excellent results in both training time and the number of model parameters.

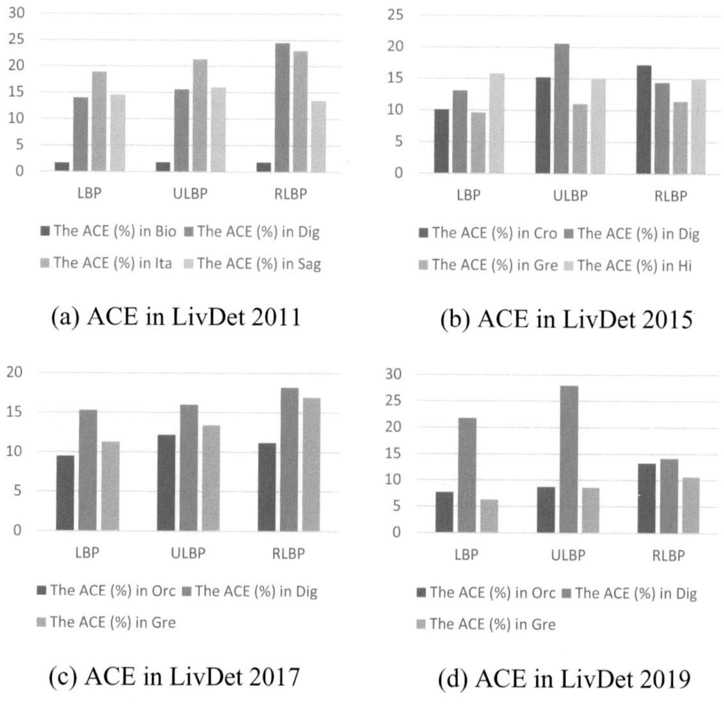

Fig. 21 The ACE of three feature descriptors. *Source: Author.*

6. Conclusion and further work

Fraudulent attack (FA) using forged fingerprints is among the most prominent security challenges for Fingerprint Authentication systems (FAS). The capacity to identify fake fingerprints at a low cost is a precondition for proper functioning and satisfactory results of FAS. Due to its shorter training time and fewer parameters, BLS can enhance the performance and applicability of FLD, especially in mobile communication devices. This paper explored applying BLS in FLD, the first-time attempt in this field. We propose a lightweight and real-time FLD method based on a comprehensive learning system with Uniform Local Binary Pattern (ULBP) for FAS. The ULBP feature descriptor represents the texture information of fingerprints, and the extracted features are fed into the BLS. Instead of inputting the fingerprint image into the BLS, the ULBP feature descriptor enables a solution with fewer dimensions. As a result, the training time of the model and parameters involved are substantively reduced while FLD performance is maintained.

The performance of existing FLD algorithms usually depends on specific training libraries. However, in the real world, algorithms must resolve the challenges of the constant emergence of new imitation materials. It is a critical problem that we need to address.

Furthermore, since fingerprint images contain users' privacy, it is difficult to collect many fingerprint samples. Therefore, enhancing the detection performance of the FLD with a small dataset of sample fingerprints is another bottleneck to breakthrough in future work.

Key terminology and definitions

Fingerprint Authentication Technology (FAT) Fingerprint authentication technology uses a person's unique fingerprint to verify their identity. It's one of the most widespread biometric authentication technologies and is used to secure everything from mobile devices to automobiles and even buildings.

Fingerprint Matching (FM) Fingerprint matching is to use algorithms to find representative features in fingerprint images and verify the authenticity of the user's identity by comparing these features. A fingerprint matching algorithm compares two given fingerprints and returns either a degree of similarity (without loss of generality, a score between 0 and 1) or a binary decision (mated/non-mated). Only a few matching algorithms operate directly on grayscale fingerprint images; most of them require that an intermediate fingerprint representation be derived through a feature extraction stage.

Fingerprint Liveness Detection (FLD) Biometrics has been an expanding industry in recent years and provides security through identifying a person based on physiological or behavior characteristics. However, it has been shown that biometric systems are vulnerable to spoof attacks by artificial fingerprint casts made of materials such as PlayDoh, silicone, or latex. There have been numerous methods proposed to solve the susceptibility of fingerprint devices to attacks by spoof fingers. One primary countermeasure to spoofing attacks is Fingerprint Liveness Detection. Liveness detection is based on the principle that additional information can be garnered above and beyond the data procured by a standard verification system, and this additional data can be used to verify if an image is authentic. Liveness detection uses either a hardware-based system or software-based system coupled with the authentication program to provide additional security. Hardware-based systems use additional sensors to gain measurements outside of the fingerprint image itself to detect liveness. Software-based systems use image processing algorithms to gather information directly from the collected fingerprint to detect liveness. These systems classify images as either live or fake.

Convolutional Neural Network (CNN) CNN is a class of artificial neural network (ANN), most commonly applied to analyze visual imagery. It is a Deep Learning algorithm which can take in an input image, assign importance (learnable weights and biases) to various aspects/objects in the image and be able to differentiate one from the other. The pre-processing required in a CNN is much lower as compared to other classification algorithms. While in primitive methods filters are hand-engineered, with enough training, CNN have the ability to learn these filters/characteristics. The architecture of a CNN is analogous to that of the connectivity pattern of Neurons in the Human Brain and was inspired by the organization of the Visual Cortex. Individual neurons respond to stimuli only in a restricted region of the visual field known as the Receptive Field. A collection of such fields overlaps to cover the entire visual area.

Broad Learning System (BLS) BLS is a paradigm shift in discriminative learning and a fast and accurate learning without deep structure. The BLS utilizes the power of incremental learning. That is without stacking the layer-structure, the designed neural networks expand the neural nodes broadly and update the weights of the neural network incrementally when additional nodes are needed and when the input data entering to the neural networks continuously. The designed network structure and incremental learning algorithm are suitable for modeling and learning big data environment. Experiments indicate that the designed structure and algorithm out-perform existing structures and learning algorithms. Uniform Local Binary Pattern (LBP) LBP is a simple and efficient texture operator which labels the pixels of an image by thresholding the neighborhood of each pixel and considers the result as a binary number. Due to its discriminative power and computational simplicity, LBP texture operator has become a popular approach in various applications. It can be seen as a unifying approach to the traditionally divergent statistical and structural models of texture analysis. The most important property of the LBP operator in real-world applications is its robustness to monotonic gray-scale changes caused, for example, by illumination variations. Another important property is its computational simplicity, which makes it possible to analyze images in challenging real-time settings.

False Accept Rate (FAR) FAR is the probability that an authentic fingerprint being wrongly identified as fake fingerprint.

False Reject Rate (FRR) FRR is the probability of a fake fingerprint being wrongly identified as an authentic fingerprint.

References

[1] C. Yin, J. Xi, R. Sun, J. Wang, Location privacy protection based on differential privacy strategy for big data in industrial internet of things, IEEE Trans. Industr. Inform. 14 (8) (2018) 3628–3636.
[2] C. Yuan, Z. Xia, X. Sun, D. Sun, R. Lv, Fingerprint liveness detection using gray level co-occurrence matrix based texture feature, Int. J. Grid Distrib. Comput. 9 (10) (2016) 65–78.
[3] China Industry Study Web, China Biometric Technology Market Research and Development Trend Analysis Report, 2021. https://www.cir.cn/6/01/ShengWuShiBieFaZhanQuShiYuCeFenX.html.
[4] M. Downs, Handbook of fingerprint recognition, JFI 54 (6) (2004) 719.
[5] Z. Akhtar, C. Micheloni, G.L. Foresti, Biometric liveness detection: challenges and research opportunities, IEEE Secur. Priv. 13 (5) (2015) 63–72.
[6] C. Yuan, X. Sun, Q.M.J. Wu, Difference co-occurrence matrix using BP neural network for fingerprint liveness detection, Soft Comput. 23 (13) (2019) 5157–5169.
[7] P.S. Prasad, B.S. Devi, M.J. Reddy, V.K. Gunjan, A survey of fingerprint recognition systems and their applications, in: ICCCE 2018: Proceedings of the International Conference on Communications and Cyber Physical Engineering 2018, Springer, Singapore, 2019, pp. 513–520.
[8] J.K.P.S. Yadav, Z.A. Jaffery, L. Singh, I. Recognizer, A short review on machine learning techniques used for fingerprint recognition, J. Crit. Rev. 7 (13) (2020) 2768–2773.
[9] W. Jian, Y. Zhou, H. Liu, Densely connected convolutional network optimized by genetic algorithm for fingerprint liveness detection, IEEE Access 9 (2021) 2229–2243.
[10] M. Komeili, N. Armanfard, D. Hatzinakos, Liveness detection and automatic template updating using fusion of ECG and fingerprint, IEEE Trans. Inf. Forensics Secur. 13 (7) (2018) 1810–1822.
[11] H. Jung, Y. Heo, S. Lee, Fingerprint liveness detection by a template-probe convolutional neural network, IEEE Access 7 (2019) 118986–118993.
[12] C. Yuan, Z. Xia, X. Sun, Q.M.J. Wu, Deep residual network with adaptive learning framework for fingerprint liveness detection, IEEE Trans. Cogn. Develop. Syst. 12 (3) (2020) 461–473.
[13] T. Chugh, A.K. Jain, Fingerprint spoof detector generalization, IEEE Trans. Inf. Forensics Secur. 16 (2020) 42–55.
[14] F. Liu, G. Liu, Q. Zhao, L. Shen, Robust and high-security fingerprint recognition system using optical coherence tomography, Neurocomputing 402 (2020) 14–28.
[15] E. Marasco, A. Ross, A survey on antispoofing schemes for fingerprint recognition systems, ACM Comput. Surv. 47 (2) (2014) 1–36.
[16] H. Kang, B. Lee, H. Kim, et al., A study on performance evaluation of the liveness detection for various fingerprint sensor modules, in: Knowledge-Based Intelligent Information & Engineering Systems, International Conference, Kes, Oxford, UK, September 2003, pp. 1245–1253.
[17] S. Schuckers, A. Abhyankar, Detecting liveness in fingerprint scanners using wavelets: results of the test dataset, in: Biometric Authentication, Eccv International Workshop, Bioaw, Prague, Czech Republic, May 2004, pp. 100–110.
[18] L. Ghiani, D.A. Yambay, et al., Review of the fingerprint liveness detection (LivDet) competition series: 2009 to 2015, Image Vis. Comput. 58 (2016) 110–128.
[19] G.L. Marcialis, A. Lewicke, B. Tan, et al., First international fingerprint liveness detection competition—LivDet 2009, in: International Conference on Image Analysis and Processing, Springer, Berlin, Heidelberg, 2009, pp. 12–23.
[20] D. Yambay, L. Ghiani, P. Denti, et al., LivDet 2011—fingerprint liveness detection competition 2011, in: Iapr International Conference on Biometrics, 2012, pp. 208–215.
[21] L. Ghiani, D. Yambay, V. Mura, et al., LivDet 2013 fingerprint liveness detection competition 2013, in: Iapr International Conference on Biometrics, 2013, pp. 1–6.

[22] V. Mura, L. Ghiani, G.L. Marcialis, F. Roli, D.A. Yambay, S.A. Schuckers, LivDet 2015 fingerprint Liveness Detection Competition 2015, in: proc, IEEE 7th international conference on biometrics, September 2015, pp. 1–6.
[23] V. Mura, G. Orru, R. Casula, et al., LivDet 2017 fingerprint liveness detection competition 2017, in: 2018 International Conference on Biometrics (ICB), 2018, pp. 297–302.
[24] A. Antonelli, R. Cappelli, D. Maio, et al., A new approach to fake finger detection based on skin distortion, in: Advances in Biometrics: International Conference, ICB 2006, Hong Kong, China, 2005, pp. 221–228.
[25] E. Park, X. Cui, W. Kim, et al., End-to-end fingerprints liveness detection using convolutional networks with gram module, arXiv (2018) preprint arXiv:1803.07830.
[26] R.F. Nogueira, R.D.A. Lotufo, R.C. Machado, Fingerprint liveness detection using convolutional neural networks, IEEE Trans. Inf. Forensics Secur. 11 (6) (2017) 1206–1213.
[27] C. Chen, Z. Liu, Broad learning system: an effective and efficient incremental learning system without the need for deep architecture, IEEE Trans. Neural Netw. Learn. Syst. 29 (1) (2017) 10–24.
[28] D. Osten, H. Carim, M. Arneson, et al. Biometric, Personal Authentication System: U.S. Patent 5,719,950, 1998.
[29] K. Karu, A.K. Jain, Fingerprint classification, Pattern Recognit. 29 (3) (1996) 389–404.
[30] A. Jain, L. Hong, R. Bolle, On-line fingerprint verification, IEEE Trans. Pattern Anal. Mach. Intell. 19 (4) (1997) 302–314.
[31] T.Y. Jea, V. Govindaraju, A minutia-based partial fingerprint recognition system, Pattern Recognit. 38 (10) (2005) 1672–1684.
[32] K. El-Khatib, L. Korba, Y. Xu, et al., Privacy and security in e-learning, Int. J. Distance Educ. Technol. 1 (4) (2003) 1–19.
[33] C.M. Medaglia, A. Serbanati, An overview of privacy and security issues in the internet of things, in: The Internet of Things: 20th Tyrrhenian Workshop on Digital Communications, Springer, New York, 2010, pp. 389–395.
[34] M. Chanson, A. Bogner, D. Bilgeri, et al., Blockchain for the IoT: privacy-preserving protection of sensor data, J. Assoc. Inf. Syst. 20 (9) (2019) 1274–1309.
[35] X. Jia, X. Yang, K. Cao, Y. Zang, N. Zhang, R. Dai, X. Zhu, J. Tian, Multi-scale local binary pattern with filters for spoof fingerprint detection, Inform. Sci. 268 (2014) 91–102.
[36] G.L. Marcialis, F. Roli, A. Tidu, Analysis of fingerprint pores for vitality detection, in: Pattern Recognition (ICPR), 2010 20th International Conference on, IEEE, 2010, pp. 1289–1292.
[37] F. Liu, G. Liu, X. Wang, High-accurate and robust fingerprint anti-spoofing system using optical coherence tomography, Expert Syst. Appl. 130 (15) (2019) 31–44.
[38] A.A. Alshdadi, R. Mehboobb, H. Dawood, M.O. Alassafi, R. Alghamdi, H. Dawooda, Exploiting level 1 and level 3 features of fingerprints for liveness detection, Biomed. Signal Process. Control 61 (2020) 1–14.
[39] Z. Xia, C. Yuan, R. Lv, X. Sun, N. Xiong, Y. Shi, A novel weber local binary descriptor for fingerprint liveness detection, IEEE Trans. Syst. Man Cybern. Syst. 50 (4) (2020) 1526–1536.
[40] Q. Zhao, D. Zhang, L. Zhang, L. Nan, High resolution partial fingerprint alignment using pore-valley descriptors, Pattern Recognit. 43 (3) (2010) 1050–1061.
[41] E. Marasco, C. Sansone, Combining perspiration- and morphology-based static features for fingerprint liveness detection, Pattern Recogn. Lett. 33 (9) (2012) 1148–1156.
[42] R.P. Sharma, S. Dey, Fingerprint liveness detection using local quality features, Vis. Comput. 35 (2019) 1393–1410.
[43] J. Galbally, F.A. Fernandez, J. Fierrez, J.O. Garcia, A high performance fingerprint liveness detection method based on quality related features, Future Gener. Comput. Syst. 28 (1) (2012) 311–321.

[44] L.F.A. Pereira, H.B. Pinheiro, G.D.C. Cavalcanti, T.I. Ren, Spatial surface coarseness analysis: technique for fingerprint spoof detection, Electron. Lett. 49 (4) (2013) 260–261.
[45] R. Agarwal, A.S. Jalal, K.V. Arya, A review on presentation attack detection system for fake fingerprint, Mod. Phys. Lett. B 34 (5) (2020) 1–26.
[46] Y.S. Moon, J.S. Chen, K.C. Chan, K.F. So, K.C. Woo, Wavelet based fingerprint liveness detection, Electron. Lett. 41 (20) (2005) 1112–1113.
[47] A. Abhyankar, S. Schuckers, Fingerprint liveness detection using local ridge frequencies and multiresolution texture analysis techniques, in: IEEE International Conference on Image Processing, 2006, pp. 321–324.
[48] S.B. Nikam, S. Agarwal, Texture and wavelet-based spoof fingerprint detection for fingerprint biometric systems, in: First International Conference on Emerging Trends in Engineering and Technology, IEEE Computer Society, 2008, pp. 675–680.
[49] B. Tan, S. Schuckers, Liveness detection for fingerprint scanners based on the statistics of wavelet signal processing, in: Conference on Computer Vision and Pattern Recognition Workshop, IEEE, 2006, pp. 1–8.
[50] D. Gragnaniello, G. Poggi, C. Sansone, L. Verdoliva, Fingerprint liveness detection based on weber local image descriptor, in: Biometric Measurements and Systems for Security and Medical Applications (BIOMS), 2013 IEEE Workshop on, IEEE, September, 2013, pp. 46–50.
[51] D. Gragnaniello, G. Poggi, C. Sansone, L. Verdoliva, Local contrast phase descriptor for fingerprint liveness detection, Pattern Recognit. 48 (4) (2015) 1050–1058.
[52] R.K. Dubey, J. Goh, V.L.L. Thing, Fingerprint liveness detection from single image using low-level features and shape analysis, IEEE Trans. Inf. Forensics Secur. 11 (7) (2016) 1461–1475.
[53] C. Gottschlich, E. Marasco, A.Y. Yang, B. Cukic, Fingerprint liveness detection based on histograms of invariant gradients, Biometrics (IJCB), in: 2014 IEEE International Joint Conference on, September 2014, pp. 1–7.
[54] K.T.K. Arun, R. Vinayakumar, V.V.V. Sajith, V. Sowmya, K.P. Soman, Convolutional neural networks for fingerprint liveness detection system, in: International Conference on Intelligent Computing and Control Systems (ICCS), IEEE, 2019, pp. 1–4.
[55] R.F. Nogueira, R.D.A. Lotufo, R.C. Machado, Evaluating software-based fingerprint liveness detection using convolutional networks and local binary patterns, in: IEEE Workshop on Biometric Measurements & Systems for Security & Medical Applications, IEEE, 2014.
[56] E. Park, W. Kim, Q. Li, J. Kim, H. Kim, Fingerprint liveness detection using CNN features of random sample patches, in: International Conference of the Biometrics Special Interest Group (BIOSIG), IEEE, 2016, pp. 1–4.
[57] H.Y. Jung, Y.S. Heo, Fingerprint liveness map construction using convolutional neural network, Electron. Lett. 54 (9) (2018) 564–566.
[58] C. Wang, K. Li, Z. Wu, Q. Zhao, A DCNN based fingerprint liveness detection algorithm with voting strategy, in: Chinese Conference on Biometric Recognition, 2015, pp. 241–249.
[59] Y. Zhang, D. Shi, X. Zhan, D. Cao, K. Zhu, Z. Li, Slim-ResCNN: a deep residual convolutional neural network for fingerprint liveness detection, IEEE Access 7 (2019) 91476–91487.
[60] C. Yuan, X. Chen, P. Yu, R. Meng, W. Cheng, Q.M.J. Wu, X. Sun, Semi-supervised stacked autoencoder-based deep hierarchical semantic feature for real-time fingerprint liveness detection, J. Real-Time Image Process. 17 (1) (2020) 55–71.
[61] C. Yuan, Z. Xia, L. Jiang, Y. Cao, Q.M.J. Wu, X. Sun, Fingerprint liveness detection using an improved CNN with image scale equalization, IEEE Access 7 (99) (2019) 26953–26966.

[62] C. Yuan, Q. Cui, X. Sun, Q.M.J. Wu, S. Wu, Fingerprint liveness detection using an improved CNN with the spatial pyramid pooling structure, Adv. Comput. 120 (2021) 157–193. Elsevier.
[63] K. Simonyan, A. Zisserman, Very deep convolutional networks for large-scale image recognition, arXiv (2014) preprint arXiv:1409.1556.
[64] A. Krizhevsky, I. Sutskever, G.E. Hinton, Imagenet classification with deep convolutional neural networks, Commun. ACM 60 (6) (2017) 84–90.
[65] K. He, X. Zhang, S. Ren, J. Sun, Deep residual learning for image recognition, in: Proceedings of the IEEE Conference on Computer Vision and Pattern Recognition, 2016, pp. 770–778.

About the authors

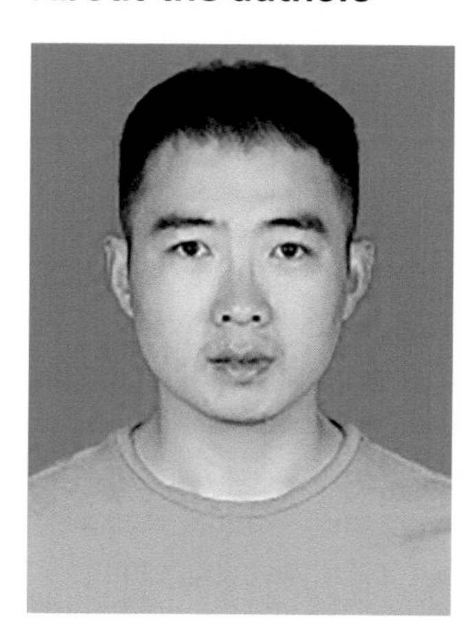

Dr. Chengsheng Yuan: Dr. Yuan is an associate professor in the School of Computer Science, Nanjing University of Information Science and Technology (NUIST), China. He obtained his B.S. and Ph.D. degree in Software Engineering from NUIST. He was a visiting student from 2017 to 2019 and a research fellow from 2019 to 2020 in the Department of Electrical and Computer Engineering at University of Windsor, Canada. His research interests include Biometric Liveness Detection, Information Hiding, Machine Learning and AI security.

Qianyue Zhang: Qianyue Zhang is pursuing her B.S. degree in Information Security at Nanjing University of Information Science and Technology, China.

Dr. Sheng Wu: Dr. Wu is a research associate at United Nations University, Operating Unit on Policy-Driven Electronic Governance (UNU-EGOV). She holds Ph.D. in Science, Technology and Innovation Management, and has extensive experience in leading research on emerging technology at national and international level.

Prof. Q.M. Jonathan Wu: Prof. Wu is a professor with in Department of Electrical and Computer Engineering, University of Windsor, Canada. He received his Ph.D. degree in electrical engineering from the University of Wales, Swansea, U.K., in 1990. He was affiliated with the National Research Council of Canada for 10 years beginning in 1995, where he became a Senior Research Officer and a Group Leader. He is a Visiting Professor in the Department of Computer Science and Engineering, Shanghai Jiao Tong University, Shanghai, China. He has published over 300 peer-reviewed papers in computer vision, image processing, intelligent systems, robotics, and integrated microsystems. His current research interests include 3-D computer vision, active video object tracking and extraction, interactive multimedia, sensor analysis and fusion, and visual sensor networks.

CHAPTER FIVE

Collaborating fog/edge computing with industry 4.0—Architecture, challenges and benefits

Arul Treesa Mathew and Prasanna Mani
SITE, VIT, Vellore, India

Contents

Advances in Computers, Volume 131
ISSN 0065-2458
https://doi.org/10.1016/bs.adcom.2023.06.001

Abstract

Edge computing is a distributed computing paradigm that takes computation and storage of data closer to the device(s) where it would be used, thus improving the response time and apparently saving bandwidth. This helped developers to create applications that could substantially reduce latencies, lower demands on network bandwidth, increase the privacy of the sensitive information, and also enable operations even when networks are disrupted. This phase shift will also help the businesses to earn more profit. Fog computing is a distributed computing scenario that is commonly used in IoT related applications. In this chapter, we discuss the essential components of Fog and Edge Computing, along with their most significant benefits.

1. Introduction to edge computing

The term 'Edge' in Edge Computing refers to keeping the computational units closer to the data source. All the cloud services like processing, real time analysis and storage of the gathered data are taken closer to the originating source. Thus processing turns faster and response time is shortened [1]. These characteristics of Edge Computing makes it widely accepted among the recent trends in computing like IoT, Virtual Reality, Real Time Response systems etc. Only the core services are sent to the Cloud tier for processing. Fig. 1 illustrates the basic architecture of Edge computing.

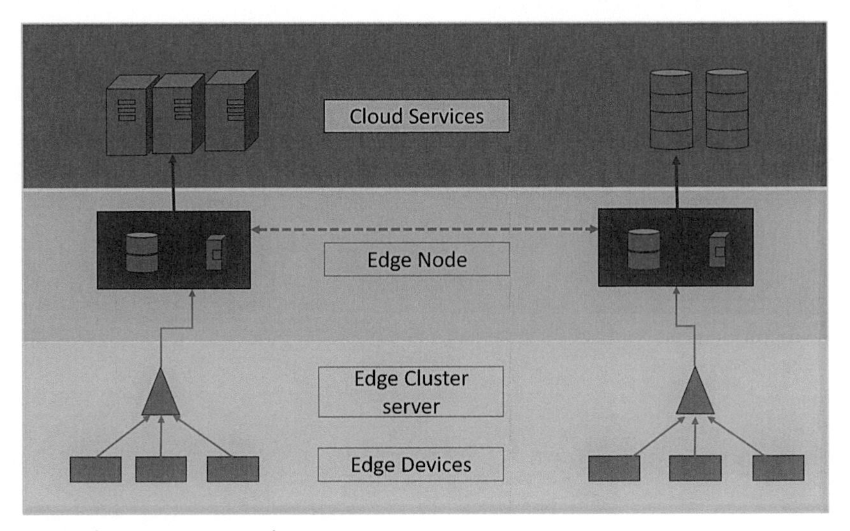

Fig. 1 Edge computing architecture.

2. The architecture of edge computing

In the subsections to follow, we will learn more about the architectural structure of Edge Computing Paradigm. The terminologies associated with edge computing, the responsibilities of each component, etc. are explained below.

2.1 Essential components

The essential components for edge computing includes a cloud, an edge device, a local edge node, edge cluster, and a gateway [2]. In the following sections, let's have a deeper look into each of these components that are essential for edge computing. Any node that is capable of performing edge computing is generally called as an edge node. Fig. 2 [3] consolidates the important components in edge computing.

2.2 Device edge

A device edge is a specialized device that is equipped with sufficient computational functions. Tasks accomplished by edge device can range from being a factory assembly machine, to even a full-fledged automobile. However, considering the economic factors, edge devices are usually found with limited computational resources. Usually, the processors associated in

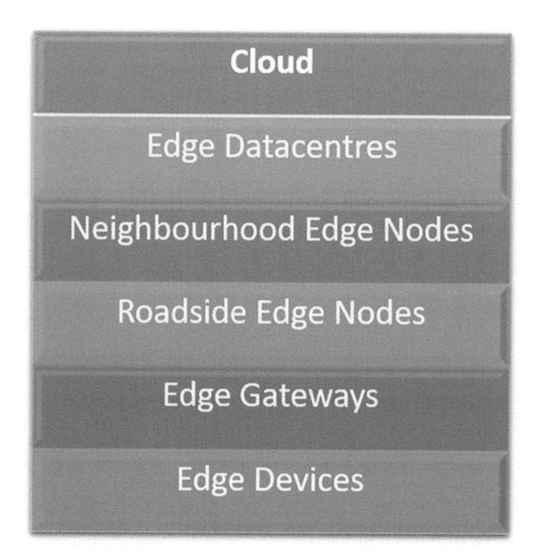

Fig. 2 Essential components.

an edge device would be either ARM or ×86 class CPUs with 1 or 2 cores. Similarly, memory capacities of edge devices would be not more than 128 MB RAM and a maximum of 1 GB of secondary storage.

2.3 Local edge

The edge closer to the source of a given data is termed as its local edge. This edge device will be in charge of processing the data, and making it available for use by the system.

2.4 Edge cluster/server

It is a general purpose computing facility that is equipped in a remote operations facility. Example for such an environment can include a factory, bank or even a retail store. The server can be built even using an industrial PC or by assembling multiple units. Computing capacity of such edge cluster servers can range from 8 to many cores, 16 GB of memory, and a local storage capacity of several hundreds of gigabytes. The role of edge cluster/server is to mainly run enterprise applications and other shared services.

2.5 Edge gateway

An edge cluster/server equipped additionally with network services is called an edge gateway. It will still host enterprise applications and shared services. In addition, network services like protocol translation, network termination, tunnelling, firewall protection, etc.

2.6 Cloud/nexus

This can be considered as one of the three layers of the edge computing architecture. The local edges forming the device layer, servers and gateways forming the network and application layer, and then the cloud layer. A cloud can be either public or private. It acts as a storage or repository for storing applications and machine learning models. It can also host and run manager applications for edge nodes.

3. The architecture of fog computing

In Fog Computing, first order operations which work on raw data are distributed to the sources. Events are detected at their origins and this could eliminate the need to store raw data in the cloud [4–8]. The operational size of the network is enhanced significantly by reducing the traffic.

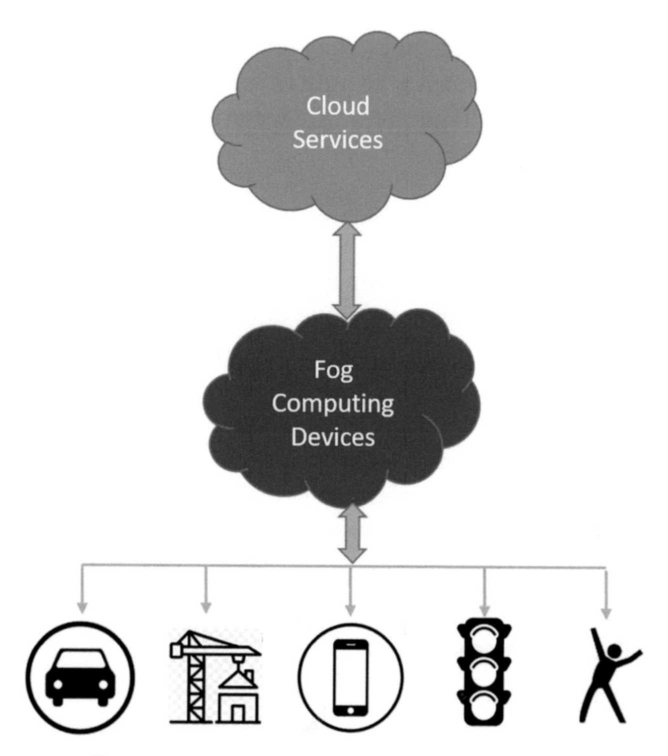

Fig. 3 Fog computing.

This advantage of Fog Computing can thus benefit large scale IoT deployments. Fig. 3 [9] illustrates the concept of Fog Computing.

Fog Computing is a new paradigm that lends itself to distribute first order operations (e.g., filtering, aggregation and translation) on raw data generated by the sources. Firstly, it allows users to detect events of interest and eliminates the need for storing raw data in the cloud. Secondly, it allows for the operational size of the network to increase many fold by reducing the total traffic. Large scale IoT deployments can benefit from leveraging the Fog Computing paradigm.

In Fog computing, there are some steps associated with the collection and processing of data. The data gathered by the end devices are first broken into chunks. These chunks are then allocated to the component nodes, after queuing them for transmission. Channels are allocated for transmission based on these queues. Idle channels get occupied with the first batch of data. The remaining chunks are allocated as and when channels are released and made available. The chunks undergo processing at the nodes and then are sorted there based on their finishing time. The processed chunks are then returned to their hosts through channels which are allocated in the same way on their first transmission. Once all the chunks are received at the host end, they are reunited to get the final data.

3.1 Essential components

Fog Computing comprises of a proper hierarchical arrangement of various physical and logical network component devices, associated hardware and dedicated software that can keep the system working well.

Essential components of Fog Computing thus include three layers—terminal layer that accommodates all the end devices, fog layer that acts as a middleware to connect end devices to the cloud, and the cloud layer which takes care of the storage and computing. In the subsections to follow, each of these layers will be explained in detail. Fig. 4 shows the hierarchical layered architecture of Fog Computing.

3.2 Terminal layer components

Terminal layer mainly include all the end devices of the IoT system. These devices communicate directly with the user and his environment. This layer

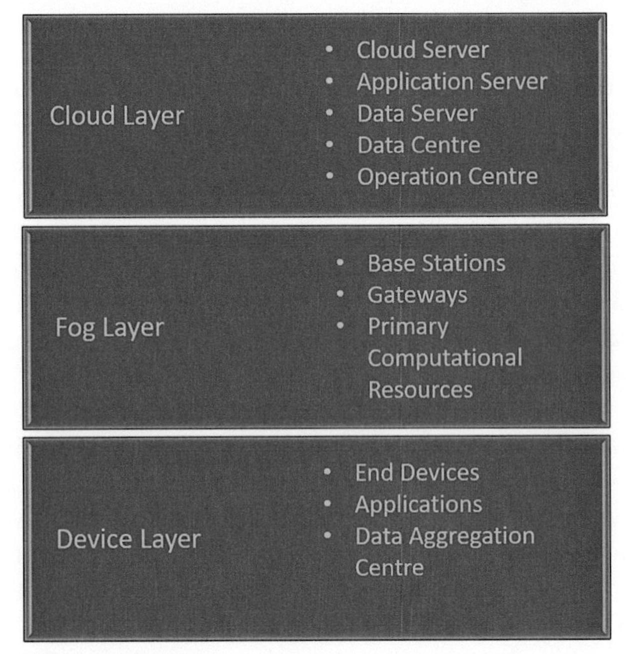

Fig. 4 Layered fog architecture.

comprises of sensors, smart cards, intelligent vehicles, readers, smart phones and other computing devices, that are used by the end user. The devices that form the terminal layer might be distributed in geographically wider locations. The main function of these devices are sensing the user environment to extract the feature data. This gathered data is then transmitted from the terminal layer to the upper layers.

3.3 Fog layer components

The second and middle layer of Fog Computing is the fog layer, which is located along the edge devices. The components of this layer is generally called as Fog nodes. These nodes include networking interfaces like routers, gateways, switches, base stations, dedicated fog servers, access points etc. As mentioned before, the main functionality of this layer is the interconnection between terminal layer end devices operating at the user environment and the cloud layer. So there are fog nodes at both the physical device end as well as the cloud end of the computing environment. These nodes can be deployed at user end like shopping malls, bus stations, commercial streets, individual retail shops, parks etc. and also at the major data centers where a cloud server would be offering a service from. These devices can either be static and fixed at a particular location or can be on motion in a mobile carrier, according to the application. The services offered from the cloud reaches the end devices through these fog nodes. The nodes offer a first level computing where the real time analysis over the gathered data from the terminal devices is done. In addition, fog nodes help in transmission of data between the top and bottom layers, and also for a short term storage of the data if necessary. The nodes are connected to the cloud system using the IP network, which assigns dedicated identity to each of these nodes in the fog layer. The collaboration between fog and cloud layers supports the back-to-back services and functionalities of the computing system. This collaboration includes data processing, storage and transmission of data.

3.4 Cloud layer components

In the top layer of the hierarchical Fog Computing Architecture, one can see many high-computing servers and large capacity storage devices. These devices help in offering the cloud services for smart networks. The data gathered by terminal devices, after a first level processing at the fog nodes, reach the cloud layer either for an advanced computing or for storage. The large computing and storage capacity of the cloud layer provides extensive computation and permanent room for the gathered data. Control strategies are

COMPONENTS IN LAYERED ARCHITECTURE

Transport Layer	→	Uploads Pre-processed and encrypted data to the cloud
Security Layer	→	Security tasks like Encryption, Decryption, Privacy and Integrity Enforcements
Temporary Storage Layer	→	Data services like distribution, replication and re-duplication, Storage Space Virtualization and Storage
Pre-processing Layer	→	Analysis, filtering, reconstruction and trimming of data
Monitoring Layer	→	Monitoring activities, resources, power, response and services
Physical and Virtualization Layer	→	Physical sensors and wireless sensor networks, virtual sensors and virtual sensor networks

Fig. 5 Components in layered architecture.

employed in the cloud layer for the efficient utilization of its functionalities and resources.

The Hierarchical Fog Architecture offers either wired or wireless connection between the fog nodes and end devices on terminal layer. Usually low power wireless access modes are utilized for wireless connectivity, like WiFi, 3G/4G/5G, Bluetooth, ZigBee etc. Whereas, they are connected to cloud layer using the core IP network. The hierarchical structure of Fog Computing helps in providing support for IoT, Cyber-Physical systems and mobile Internet through its data processing and storage capabilities.

Fig. 5 [10] describes the tier-wise components that form the layers of the Fog computing paradigm. This is adopted from the work of Atlam *et al.* [10]. Here, the transport and security layers are in the cloud side, temporary storage, pre-processing, and monitoring are with fog nodes, and the last one is the terminal layer, where the sensing and other physical devices are deployed.

4. Benefits and challenges of edge computing

Edge computing is a computing paradigm where data gets processed and analyzed in the close proximity of the originating source. This approach eliminates the efforts to transmit data to a cloud or a data center for processing. Thus the workload of the server and the traffic through the network can be reduced significantly. Real time processing of sensed data near the source as well as faster response time are the main benefits of using edge computing, thus making it suitable for IoT and Cyber Physical Systems. Edge computing paradigm can also be employed with machine learning and artificial intelligence approaches, as it can process data on a faster pace.

There are a few other benefits for adopting edge computing. But these also introduce some challenges as well. Let's see each one of them in detail. Distribution of workload and application and analytics services is the major change that edge computing has brought in. We are no longer persistently requesting the cloud for services and storage. Only the most important services are kept with the cloud and all others are distributed in locations closer to the source edge devices. This shift helps in reducing traffic inside the network. Even though this is an advantage of edge computing, along with it comes the overhead of managing this change, by using proper orchestration and automation tools, without affecting the system architecture. This means the local storage and processing units should easily be accessible and also offering real time responses so that the performance of the system is consistent. This is one challenge while using edge computing paradigm. The technical support teams have to be trained with the new tools and orchestration adopted. They have to get familiarized with the newly adopted methods and tools. Along with that, there should be a better distribution of workloads in the localized environments, so that the system becomes more portable.

Another advantage is that in edge computing, size of the system would not become a concern. Scaling the system is enhanced by the introduction of edge computing in the network. When the tools required to process and analyze the data are taken closer to its source, latency in the system gets reduced, as well as the resources required for computing at each distributed location also can be reduced. But, in this context, the distribution of workload becomes a bottleneck as the size of workloads at various local computing nodes might limit the potential of edge computing. Adopting a proper standard for the portability and other methodologies adopted, is a challenging task, as they have to deal with varying workloads. The adopted standards should also be capable to accommodate the entire application lifecycle.

If proper care is given in workload distribution, the performance of the system can be enhanced significantly. For this, workloads need to be broken into smaller pieces and distributed among the edge devices, finally consolidating the computation results at the edge cluster/server. This distribution of workloads after dividing them into smaller parts, also require migration of network and other application service components to the local processing components. Proper care, effort and expertise is thus required to deploy these components effectively so that the expected performance can be achieved. If the workloads can be prioritized, this challenge can be mitigated in a reasonable way. The factors for prioritizing workloads can range from migration benefits, complexity of partitioning and processing to the time required for migrating the workload. Studies are happening in this area to identify the migration capabilities that can be deployed for container based approaches.

One prime benefit offered in edge computing paradigm is enhanced data security. Availability of data is limited to the areas closer to its origin and not to the entire network, thus stealing the data would now require proper knowledge about the source and processing edge nodes. Given this as a benefit of edge computing, there arises a challenge to the overall security of data. Unlike in pure cloud based scenarios, the local edges have limited resources and computing capabilities and would not be able to effectively withstand the possible hazards. This demands a special attention from the network designers to make the distributed architecture capable of ensuring its overall security. By incorporating additional security mechanisms, this can be achieved, but the process and efforts become more complex as the security as well as security concerns are also distributed in edge computing.

The most important challenge in designing an edge based network is the rapid changes and innovations of device and security standards that are inevitable in the modern ecosystem. Maintaining a long term decision on the underlying technology is practically impossible due to this faster pace of upgrades that happen in technology. Different standards and technology among competing devices creates a need for a never ending analysis and innovation or upgrades. However, usage of efficient tools in managing workloads throughout the application lifecycle can also support introduction of new technology or devices to the existing system without forcing the complete redesign of the architecture.

5. Benefits and challenges of fog computing

The decentralized computing approach of Fog computing has helped in reducing the processing time at the cloud end as the first level processing and real time analysis of data occurs in the Fog nodes. This helps in achieving a better response time better bandwidth utilization. Benefits of Fog computing can be listed as reduced latency, enhanced security, improved customer experience, etc. Since application of Fog computing is mostly in IoT systems where direct user interactions are more, a better response time will help in increased customer satisfaction levels. Also, since fog computing offers real time analysis of sensed data near to its original source, any anomalies or misbehaviors can be easily detected and the user security can thus be enhanced [9–13]. This has made Fog Computing widely accepted among various businesses.

The performance measurements of using fog computing based on latency results, throughput results, and network usage results show that, if the synchronization size is kept small, latency gets reduced significantly, thus providing a better throughput and low utilization of network resources.

Along with all these above mentioned benefits of Fog Computing, there are some challenges which require a wider attention from the researchers and architecture designers. The main challenge that one could foresee is the successful enforcement of security, both to the data and the network. As fog nodes takes charge of processing and real time analysis of sensed data, how effectively can the data security be achieved in this layer of the architecture is of extreme importance. It becomes the responsibility of the layer designer to ensure that efficient devices are being deployed and proper security mechanisms are made available especially to the trustworthy nodes that carry out processing of sensitive data.

Another concern in fog computing is in the task scheduling. Since fog computing is heterogeneous in nature, the conventional scheduling would not work as the resources at each node are different, as well as the high priority tasks at each node are also different. Hence, scheduling is a challenging job in the fog computing scenario. The need of a dedicated algorithm to take care of task scheduling in this distributed and heterogeneous environment has thus become a topic of research. Similarly, choosing the right node to execute a given task is also a consideration, unlike the earlier networks. The heterogeneity that exists in the fog layer devices are to be addressed to know which device can efficiently perform a given task. This also adds to the burden of the network designer.

The nodes in a Fog network are heterogeneous in nature, i.e., there is no guarantee that the resources available with each fog node would be similar. Products from different manufacturers may vary in their processor, memory capabilities. There are chances of random mixtures of resources in the fog nodes that are used in a network. So, ensuring the interoperability and compatibility amongst this heterogeneous environment is a challenging and important task while designing the network.

Another important challenge in designing a fog network is tackling the user's mobility. Managing the data during this phase needs proper addressing as the fog nodes which took charge of data processing, storage and real time analysis will now change as the user moves to another location. Since there is no global storage space for all data contained in the network, migrating the data across the network needs special attention as this might lead to a possible door for data loss.

Scalability of the system will also become a challenge especially when fog computing is applied for IoT applications. The need of resources, bandwidth and associated network resources, restricts the growth of network beyond certain limits. Efficiency of the fog servers directly affects scalability. Once this is ensured, system can be extended to support more devices. However, this growth also makes a directly proportional impact on the network complexity and network utilization.

Dynamicity has to be ensured while designing a fog based network. The fog nodes being deployed to the system should have the capability to automatically reconfigure their arrangement and resources so as to adapt with the changes in network. It also should have proper intelligence to detect any alarming situations from the real time analysis of sensed data and also to initiate quick responses.

6. Fog computing and industry 4.0

The 4th Industrial Revolution was triggered with the wide spread usage of Information Technology and Digitization [14]. Industry 4.0 aims to provide robust, efficient industrial solutions which ensures high productivity of high quality products in a low cost. Fourth industrial revolution exhibited extensive use of ubiquitous computing and mobile Internet, along with close interactions of machine learning, artificial intelligence, and cyber physical systems in the industrial domain. Scope of industry 4.0 extends to Smart Connected Machines, Smart Factories, Gene Sequencing, Nanotechnology, Renewables, Quantum Computing etc.

The fourth industrial revolution aims to revolutionize the industrial sector in a way that it reaches to the common people. Objectives of industry 4.0 includes, providing robust solutions to real- life industrial problems, higher customer satisfaction, higher productivity, better performance and global connectivity of the industries. It aims to achieve greater production by optimizing the production decisions, with higher efficiency and availability. Industry 4.0 offers deeper level of analytics and prediction, with higher precision and accuracy in the quality of the product.

Challenges in implementing industry 4.0 includes handling time-sensitive data, need of immediate and quick response, hazardous consequences if a real-time response awaited is delayed, bottlenecks in using centralized approach etc. Since industry 4.0 tries to address real-time issues, it demands an in-time analysis and quick decision making capabilities. This enforces the need of a distributed working architecture for industry 4.0 based systems.

Fog computing thus emerged as a solution for all these challenges. Fog computing ensures distributed analysis and processing within a short time, safeguarding the security of the system. By adding an intermediate layer between edge and cloud layers, fog computing ensures increased scalability with reduced traffic. It also helps in identifying the useful data from the sensed data chunks, thus reducing the amount of raw data sent to the cloud. The intelligent devices capable of computing, known as fog nodes, are deployed in the fog layer to perform a first-level analysis of the sensed data. The fog nodes filter useful data from the sensed data, aggregate them together and translate them for being further processed at the cloud end.

6.1 Fog computing and industrial Internet of things

Industrial Internet is a collective collaboration of machines, people, computers and Internet of things. It offers intelligent and advanced analytical tools to yield better business results.

Industrial IoT enabled with Fog computing helps in addressing the existing weaknesses of industrial automation, with better process control analytics. It also provides new functionalities with additional features, with an additional enrichment to the existing ones. Fog computing offers analysis of machine, process and data in the industrial scenario. It also supports advanced and optimized modes of decision making capabilities thus helping in developing intelligent operations. It supports the edge level algorithms for real-time control, and also ensures proper filtration of usable data from the edge level, thus preventing crowding of data at the cloud end. Thus fog based IIoT offers reduced cost, better scalability, novel trading ideas, real-time monitoring and visualization, and enhanced flexibility. It affirms that the security is intact at both ends.

7. Edge computing and industry 4.0

Internet of Things and related technologies increased the number of connected devices to an exponential extent, in all areas of its applications. Apparently, industrial sector also witnessed a radical change with the introduction of IoT based solutions. This advancement resulted in generating an exponential amount of data collectively called the "Big Data." This huge data needs better processing and analytical capabilities which cannot be addressed by the conventional cloud computing methods and will lead to high latency and delayed response. However, this cannot be encouraged especially in the case of real-time systems that expect timely responses.

One solution to this issue is to carry out the computations closer to the device which senses and generates the data. Edge computing based technologies become handy in this scenario. The advanced and efficient modes of operations offered by industry 4.0 results in higher amounts of data, which if incorporated with edge based solutions will help in providing efficient and effective solutions to real world industrial problems. Edge computing overrides cloud based solutions by offering a real-time analysis of the sensed data at a location closer to the data source, thus offering distributed processing and analysis of the data within a short time. Whereas, cloud based solutions end up in high latency and delayed response in this aspect. Cloud computing technologies become more useful for those applications which emphasizes on storing the collected data. Edge computing solutions fail in this aspect due to their poor storage capacities. Still, they offer distributed storage solutions which help in keeping the sensed data as close as possible to its source [15–18].

Applying edge computing to industrial processes will use the computing capacities of intelligent devices and equip the system with self-decision-making capacities on the analysis of sensed data. As a result, this will lead to real-time responses in the manufacturing processes.

Storage infrastructure [18] in edge computing has to be enhanced in all levels like ephemeral at the lowest level to semi-permanent at the highest, spanning the geographical are for a longer period of time. However, this might lead to an additional and significantly higher amount of expense, which thus making the situation favorable for cloud based storage solutions. Thus the industry 4.0 looks into edge based solutions for real time response and temporary storage of the analyzed data, and into cloud based solutions for a persistent storage. Edge based solutions also offer a higher level of security to the system compared to cloud based solutions as there is a fine level of first hand processing of data very close to the source.

7.1 Cloud computing v/s edge computing in industry 4.0

In the context of industry 4.0, cloud based solutions are found vulnerable in the network security and data security aspects as the storage is made available outside the industrial environment. Whereas, both these concerns are addressed in the edge based solutions as they are processing and storing the sensed data as close as possible to its source.

Another major challenge for cloud computing in industry 4.0 is that there is an increase in the time required for processing and response. This occurs due to two reasons—geographical distance of the processing and computing facilities from the data source, and the huge amount of data that gets accumulated at these facilities from various sources. On the contrary, edge computing offers a reduced processing and response time owing to the fact that it performs distributed processing and analysis of the data keeping the facilities in the close proximity of the source. Network delay is significantly low in edge based solutions compared to that of the cloud based ones.

Cloud computing methods are more expensive compared to the edge computing models, due to the need of highly sophisticated storage and processing solutions to handle Big Data. Transfer of the analysis and findings may also incur an additional expense if the same is done in a paid facility. Whereas, Edge Computing methods are found more affordable as they use less-expensive IoT devices and do not require any additional expenses for special services, such as analysis transfer.

Fig. 6 shows the architectural framework of edge computing with respect to cloud computing.

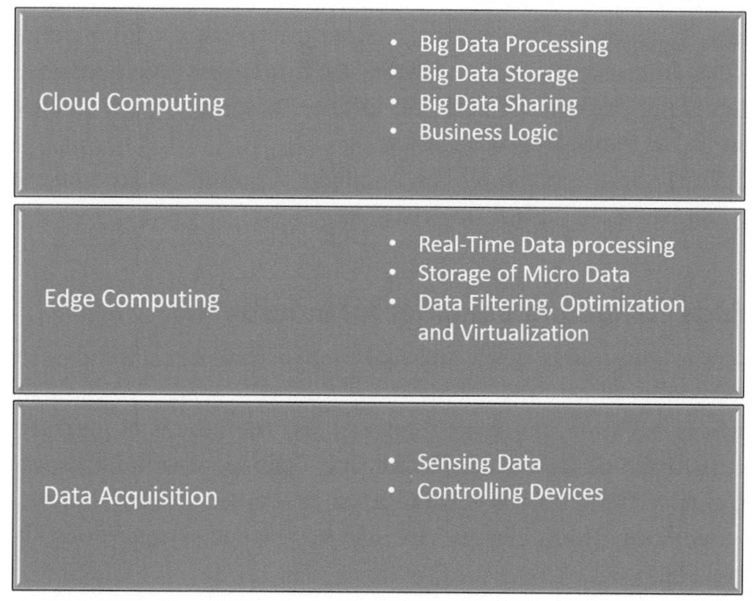

Fig. 6 The framework of differences between edge computing and cloud computing.

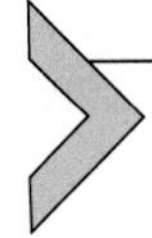

8. Use cases of fog/edge computing in industry 4.0

8.1 Fog computing use cases for industry 4.0

There are many industries in which fog based IIoT solutions have brought in efficient solutions, like mining, transportation, smart grid, oil and gas industry etc. Risky work environments like mines are now monitored remotely with the help of sensors and other intelligent devices, thus saving the human life and time. Also, such fog based solutions are cost-effective, scalable, flexible, and they also ensure reduced processing time and increased productivity.

In mines, fog based IIoT solutions help in remote monitoring of risky environments, achieving enhanced productivity and reduced operational cost, predictive analysis of operational failures, and increased accuracy through the real time analysis of sensed data by the fog nodes. In the case of oil and gas industries fog computing IIoT acts in offering advanced real-time operations, helps in monitoring the system thereby detecting any unusual events, step-by-step automation of the conventional processes and machinery, scalable and adaptable solutions, and in real time computation, control and management of resources and data.

Smart grid industry is benefitted using fog based IIoT alternatives as, it aids to meet the dynamic power demands of the appliances, ensures bidirectional communication among producers using advanced metering infrastructure, suppliers and consumers, proper power supply from the micro grids, etc. In transportation, fog based IIoT solutions offers services like smart parking, Internet of vehicles, location accuracy, smart traffic-light systems, other location-aware services etc.

FogHorn, Sonm, Nebbiolo Technologies, Crosser etc. are some of the major Fog based IIoT platform providers.

8.2 Edge computing use cases for industry 4.0

Edge computing offers quick and real-time analysis of sensed data in the close proximity of its source. This makes it more accepted for those industries where real time monitoring, analysis and response is of great importance. Such industries include healthcare, defense, weather forecast etc. where a delayed response is equal to a severe disaster. In addition, the data and network security is best preserved in edge based systems thus offering more reliability.

An example of edge computing based solution in healthcare sector is described in the work of Pace *et al.* They have devised a solution called

BodyEdge, which consists of a tiny mobile client module and a performing edge gateway that supports multi-radio and multi-technology communication in order to collect and locally process data coming from various deployment scenarios. This approach is best suited for human-centric applications.

Edge computing based IIoT technologies are used for Condition-based Maintenance, Real-Time Safety Monitoring, Facial Recognition Systems etc. The assembly line sensors are used to detect anomalies or other maintenance issues spanning from forklifts to aircrafts. This will help in reducing the costs associated with repairs, replacements, and downtime. Edge based IIoT devices helps to provide safer working environment in the organizations even in remote locations. Risky work areas like mines, oil rigs etc., now employ edge based technologies to monitor the works associated with dangerous equipment. Edge based solutions can also be employed in security monitoring to detect any potential threats and to raise alerts.

INTELLIEDGE by Fujitsu Global is an example of edge computing based IIoT solution for industry 4.0.

9. Conclusion

In this chapter, we have discussed the architecture of edge computing and fog computing. Edge computing is an advanced computing paradigm that offers to process data close to its originating source so that real time analysis of the same can be performed at the earliest. Essential components of edge computing include various edge devices, edge gateways, edge servers, and a cloud. Edge computing has resulted in faster response times in the system, thus making it applicable on advanced computing systems like IoT, machine learning, artificial intelligence etc. The chapter also discussed about Fog Computing Architecture. The commonly adopted Hierarchical model of Fog Computing was explained in the chapter. The three layers of the architecture and their roles were discussed. Fog nodes helps in reducing the workload of the cloud infrastructure as they take care of the first level processing and real time analysis of sensed data. This method helps in reducing the latency of the system and also in enhancing the security. Challenges and benefits of both edge and fog computing were listed out in this chapter. The chapter also gives an insight to the close association of fog/edge based computing solutions in industry 4.0, advantages of these technologies over the cloud based technologies, and also familiarizes some use cases of these computing models in various industrial applications.

References

[1] W.Z. Khan, E. Ahmed, S. Hakak, I. Yaqoob, A. Ahmed, Edge computing: a survey, Futur. Gener. Comput. Syst. 97 (2019) 219–235, https://doi.org/10.1016/j.future.2019.02.050.
[2] Edge Computing Architecture and Use Cases: Benefits, Challenges, and Architectures When Enterprises Implement Edge Computing in Telecom and Other Industries, IBM Developer, 7 April 2022. https://developer.ibm.com/depmodels/edge-computing/articles/edge-computing-architecture-and-use-cases.
[3] The Edge Is Near: An Introduction to Edge Computing! Inovex GmbH, 2019. Retrieved November 1, 2021, from https://www.inovex.de/de/blog/edge-computing-introduction/.
[4] V. Gazis, A. Leonardi, K. Mathioudakis, K. Sasloglou, P. Kikiras, R. Sudhaakar, Components of fog computing in an industrial internet of things context, in: 2015 12th Annual IEEE International Conference on Sensing, Communication, and Networking - Workshops (SECON Workshops), Seattle, WA, USA, 2015, pp. 1–6. https://doi.org/10.1109/SECONW.2015.7328144.
[5] Y. Liu, J.E. Fieldsend, G. Min, A framework of fog computing: architecture, challenges, and optimization, IEEE Access 5 (2017) 25445–25454, https://doi.org/10.1109/ACCESS.2017.2766923.
[6] O. Osanaiye, S. Chen, Z. Yan, R. Lu, K.-K. R. Choo, M. Dlodlo, From Cloud to Fog Computing : A Review and a Conceptual Live VM Migration Framework, IEEE Access, 2017, pp. 8284–8300.
[7] P. Hu, S. Dhelim, H. Ning, T. Qiu, Survey on fog computing: architecture, key technologies, applications and open issues, J. Netw. Comput. Appl. 98 (2017) 27–42, https://doi.org/10.1016/j.jnca.2017.09.002.
[8] M.R. Anawar, S. Wang, M.A. Zia, A.K. Jadoon, U. Akram, S. Raza, Fog computing: An overview of big IoT data analytics, Wirel. Commun. Mob. Comput. 2018 (2018) 7157192. https://doi.org/10.1155/2018/7157192.
[9] B. Posey, S. Shea, I. Wigmore, What Is Fog Computing?, IoT Agenda, 2020. https://internetofthingsagenda.techtarget.com/definition/fog-computing-fogging.
[10] H. Atlam, R. Walters, amp; Wills, G., Fog computing and the internet of things: a review, Big Data Cogn. Comput. 2 (2) (2018) 10, https://doi.org/10.3390/bdcc2020010.
[11] N. Joshi, A. Bouargane, Benefits of Fog Computing, BBN Times, 17 December 2020. https://www.bbntimes.com/technology/benefits-of-fog-computing.
[12] I.U. Din, M. Guizani, S. Hassan, B.-S. Kim, K. Khan, M. Atiquzzaman, S.H. Ahmed, The Internet of things: a review of enabled technologies and future challenges, IEEE Access 1 (2018) 7606–7640, https://doi.org/10.1109/ACCESS.2018.2886601.
[13] Z. Hao, E. Novak, S. Yi, Q. Li, Challenges and software architecture for fog computing, IEEE Internet Comput. 21 (2) (2017) 44–53, https://doi.org/10.1109/MIC.2017.26.
[14] National Programme on Technology Enhanced Learning, Introduction to Industry 4.0 and Industrial Internet of Things, Course, 2021. Retrieved November 1, 2021, from https://onlinecourses.nptel.ac.in/noc21_cs66/preview.
[15] B. Bajic, I. Cosic, B. Katalinic, S. Moraca, M. Lazarevic, A. Rikalovic, Edge computing vs. cloud computing: challenges and opportunities in industry 4.0, in: B. Katalinic (Ed.), Proceedings of the 30th DAAAM International Symposium, DAAAM International, Vienna, Austria, 2019, pp. 0864–0871, ISBN: 978-3-902734-22-8, https://doi.org/10.2507/30th.daaam.proceedings.120.
[16] Z. Lin, J. Liu, J. Xiao, S. Zi, A survey: resource allocation technology based on edge computing in IIoT, in: 2020 International Conference on Communications, Computing, Cybersecurity, and Informatics (CCCI), 2020, pp. 1–5, https://doi.org/10.1109/CCCI49893.2020.9256663.

[17] P. Pace, G. Aloi, R. Gravina, G. Caliciuri, G. Fortino, A. Liotta, An edge-based architecture to support efficient applications for healthcare industry 4.0, IEEE Trans. Industr. Inform. 15 (1) (2019) 481–489, https://doi.org/10.1109/TII.2018.2843169.
[18] C. Yang, S. Lan, W. Shen, L. Wang, G.Q. Huang, Software-defined cloud manufacturing with edge computing for industry 4.0, in: 2020 International Wireless Communications and Mobile Computing (IWCMC), 2020, pp. 1618–1623, https://doi.org/10.1109/IWCMC48107.2020.9148467.

Further reading

[19] K. Shaw, What Is Edge Computing and Why It Matters, Network World, 2019. https://www.networkworld.com/article/3224893/what-is-edge-computing-and-how-it-s-changing-the-network.html.
[20] S. Wang, Edge computing: applications state-of-the-art and challenges, Adv. Netw. 7 (1) (2019) 8–15, https://doi.org/10.11648/j.net.20190701.12.
[21] S. Hamdan, M. Ayyash, S. Almajali, Edge-computing architectures for internet of things applications: a survey, Sensors (Basel, Switzerland) 20 (22) (2020) 6441, https://doi.org/10.3390/s20226441.
[22] What Is Edge Computing. IBM. (n.d.). https://www.ibm.com/cloud/what-is-edge-computing.
[23] I. Sittón-Candanedo, R.S. Alonso, Ó. García, L. Muñoz, S. Rodríguez-González, Edge computing, IoT and social computing in smart energy scenarios, Sensors 19 (15) (2019) 3353, https://doi.org/10.3390/s19153353.
[24] What Is Edge Computing? Advantages of Edge Computing—Alibaba Cloud Knowledge Base. (n.d.). https://www.alibabacloud.com/knowledge/what-is-edge-computing.
[25] R. O'Day, Edge Computing and Thermal Management, 2020. https://www.qats.com/cms/2020/01/14/edge-computing-and-thermal-management/.
[26] A. Mohammad, S. Zeadally, K.A. Harras, Deploying fog computing in industrial internet of things and industry 4.0, IEEE Trans. Industr. Inform. (2018), https://doi.org/10.1109/TII.2018.2855198.
[27] S. Sarkar, S. Chaterjeeand, S. Misra, Assessment of the suitability of fog computing in the context of internet of things, IEEE Trans. Cloud Comput. Secur. 6 (1) (2018) 46–59.
[28] F. Bonomi, R.A. Milito, P. Natarajan, J. Zhu, Fog computing: a platform for internet of things and analytics, in: Big Data and Internet of Things, Springer, 2014, pp. 169–186.
[29] I. Sittón-Candanedo, R.S. Alonso, S. Rodríguez-González, J.A. García Coria, F. De La Prieta, Edge computing architectures in industry 4.0: a general survey and comparison, in: Á.F. Martínez, L.A. Troncoso, M.J. Sáez, H. Quintián, E. Corchado (Eds.), 14th International Conference on Soft Computing Models in Industrial and Environmental Applications (SOCO 2019). SOCO 2019. Advances in Intelligent Systems and Computing, vol. 950, Springer, Cham, 2020, https://doi.org/10.1007/978-3-030-20055-8_12.
[30] S. Trinks, C. Felden, Edge computing architecture to support real time analytic applications: a state-of-the-art within the application area of smart factory and industry 4.0, in: 2018 IEEE International Conference on Big Data (Big Data), 2018, pp. 2930–2939, https://doi.org/10.1109/BigData.2018.8622649.
[31] Wikimedia Foundation, Edge Computing, Wikipedia, 2021. Retrieved November 1, 2021, from https://en.wikipedia.org/wiki/Edge_computing.

About the authors

Mrs. Arul Treesa Mathew is a research scholar at School of Information Technology and Engineering, VIT, Vellore, pursuing her Doctoral Research in Healthcare IoT. She has published papers in both international journals and conferences. She has also authored book chapters for reputed publishers. Her areas of interest include cybersecurity, IoT systems, and deep learning.

Dr. Prasanna Mani is a professor in the Department of Smart Computing of School of Information Technology and Engineering at Vellore Institute of Technology, Vellore campus. He is an expert in the field of Software Testing. He has published many international journal articles in the broad areas of Computer Science Engineering, Software Modeling and Testing, Internet of Things, Software Security, Requirement Engineering, Social Engineering, Decision Sciences, Intelligent Systems, etc. His articles are well cited and he has an H-Index of 10 according to the latest scopus sources. His major contributions are in Graphical User Interfaces, Model-Based Testing, Computer Interface, Unified Modeling Language, Model-Based Testing, Software, Symbolic Execution, Test Generation and Quality of Life. In his decade-long academic career, he was invited as the keynote speaker in many reputed conferences and has also delivered many guest lectures abroad. He is always passionate and enthusiastic in updating himself on the recent trends in technology and education. Under his expert supervision, three PhD scholars and one MS (by Research) scholar have been awarded degrees. In addition, he authored a book titled "Interview Psychology and Industrial Perspective on C" and has also published several book chapters with reputed publishers.

CHAPTER SIX

Virtual Raspberry Pi-s with blockchain and cybersecurity applications

Marcus Tanque[a] and Phillip Bradford[b]
[a]Independent Researcher, Washington, DC, United States
[b]Department of Computer Science and Engineering, University of Connecticut, Storrs, CT, United States

Contents

Abstract

Cybersecurity applications can run on large physical servers. These applications can run on small and constrained virtual Raspberry Pi-s. Is there any advantage to virtualizing Raspberry Pi-s and running cybersecurity applications on these small virtual devices?

Advances in Computers, Volume 131
ISSN 0065-2458
https://doi.org/10.1016/bs.adcom.2023.04.005

In trying to answer the question, this chapter explores cybersecurity applications on virtual Raspberry Pi-s augmented with blockchains. Virtual computers offer migration, redundancy, replication, cloning, provisioning, and agility. This chapter investigates whether such attributes benefit cybersecurity applications on small devices augmented with blockchains. Raspberry Pi-s are concrete examples for this paper; other tiny computing devices are suitable. The paper briefly discusses blockchains run by networks of virtual Raspberry Pi-s. Independent Raspberry Pi networks can run blockchains to host ledgers. The ledgers can irrevocably store and track events. RPi-hosted blockchains can have periodic fingerprints embedded in other sizeable public blockchain ledgers. Entire Raspberry Pi image fingerprints can be stored in blockchain ledgers. Keeping these fingerprints helps secure virtual Raspberry Pi-s, small blockchains, and records of events recorded in these small blockchain ledgers. The process allows for validating small RPi-hosted blockchains in larger, better-known ones. It may lead to challenging deepfakes since small virtual devices can follow people or events while constantly documenting their whereabouts—this process quickly challenges deepfakes.

1. Introduction

The continuous need for more data storage, cybersecurity applications, agility, and reliability can define the increasing digitalization of global computer systems. This process includes scalable technology capabilities that meet consumers' and organizations' requirements. The rise of blockchain, cybersecurity applications, virtual Raspberry Pi, and other notable industrial solutions is vital to such a technology paradigm shift in recent decades. Despite these advances, blockchain, cybersecurity, and other technologies are prone to adversarial and insider threats [1]. A Virtual Raspberry Pi is a physical Raspberry Pi computer's simulated version. The virtual Raspberry Pi computer can be installed with the help of the Raspberry Pi Operating System (OS). It provides the computer with core software functions. These functions allow a virtual machine to emulate a virtual Raspberry Pi to create instances that run without physical hardware [2]. The OS provides virtual and physical Raspberry Pi hardware and software resources they need for computing activities.

Due to cost-effectiveness, virtual Raspberry Pi-s can be used for testing and development. The QEMU or VMWare emulators can host virtual RPi-s. As a software-based emulation supporting physical Raspberry computer hardware components, Virtual Raspberry Pi allows users to create and launch virtual instances. The software-based emulation allows users to create and launch virtual Raspberry Pi virtual instances on designated computers. These activities can run on the virtual instance without requiring physical hardware components. Users can apply these techniques when developing, learning, and testing Raspberry Pi systems. The process is cost-effective and does not require much space utilization and/or physical devices to launch. Virtual Raspberry Pi can be procured and installed on various systems. Most of these systems can be acquired with OSs and applications that users need to perform basic activities. Virtual Raspberry Pi-s are examples of constrained systems that can run computer security applications. This chapter focuses on cybersecurity applications and virtual Raspberry Pi-s. Virtual Raspberry Pi-s participate in building small blockchains. Small virtual Raspberry Pis can follow people and continue to build small blockchains. Following people or events can be used to document a person's location over time to form a defense against possible deepfakes, which would presumably not have such valid records of a person's activities [3,4].

Virtual RPi-s can run in defensive and offensive security environments because they are general-purpose computers, even though they are constrained. Raspberry Pi-s can run defensive intrusion detection tools like Snort [5,6]. Offensive tools can be built from facial recognition systems using Open Source Computer Vision Library/OpenCV and Python libraries, such as face recognition, imutils, and air cracking. Cybersecurity applications and Raspberry Pi solutions can be integrated into traditional computer security. Raspberry Pi-s are small computers containing several connectivity options. These devices can be connected using Wi-Fi, Bluetooth, and Ethernet. In the virtual case, these physical connections are not directly used. Securing the Raspberry Pi-s software is fundamental to virtual RPi-s security. This chapter does not discuss securing Raspberry Pi computers at length. Physical Raspberry Pi-s are inexpensive, reliable, dependable, and agile. Virtualizing physical Raspberry Pi-s amplifies these attributes [5,7].

This chapter also focuses on applying virtual Raspberry Pi-s to cyber security and briefly examines blockchain technology. It provides a brief overview of surveillance monitoring and detection systems. These systems include but are not limited to Closed Circuit Television (CCTV), Digital Video Recording, and Internet Protocol (IP) cameras [8]. CCTV systems are video networks that are isolated from other networks. These machines add value to other security systems since they are not accessible externally. Virtual Raspberry Pi-s can control the video and perform face recognition and related security operations in isolation. The process is vital for monitoring and securing systems against intruders. It is critical to prevent security systems from being used against their owners. For instance, secured CCTV systems may be connected to open networks to report incidents periodically. These systems may have a Raspberry Pi occasionally connected to their networks to copy or build a ledger of incidents.

The Pi-hole is a messaging and tracking system that can be used as an intrusion detection system for Raspberry Pi-s. It is an open-source network tracker and an advertisement blocker. Such a system is designed to block traditional websites and Domain Name System (DNS) requests. Pi-hole can filter DNS by blocking suspicious domains. This process serves as a resource-saving feature and security mechanism. It is helpful for security since advertisements can carry malware, distract users, and gather data for social engineering [9]. The chapter discusses motion passive infrared sensor systems and how these solutions can be integrated with virtual Raspberry Pi-s. Passive Infrared (PIR) are electronic sensors that measure infrared light. These devices similarly detect motion while Raspberry Pi-s manages PIR sensors. Infrared sensors detect movement and body and any infrared energy changes.

2. Background

Virtual Raspberry Pi can be deployed as a software program to mimic the physical Raspberry Pi hardware capability. The virtual Raspberry Pi can support RaspiBlitz, Blockchain networks, and Bitcoin while providing additional security capabilities/features needed to support the environment. Over the years, RaspiBlitz has played a pivotal role in the enterprise security domain. RaspiBitz users can produce and safeguard their private keys in the virtual Raspberry Pi offline environment.

To access and transfer cryptocurrencies, users need private keys. These keys are mainly stored in the digital wallet. The downside is that if hackers can access the user's wallet, they can then pilfer the private keys and any funds stored in the digital wallet. Using the RaspiBlitz digital cryptocurrency, users can store their private keys on a virtual Raspberry Pi platform without accessing the Internet. This security measure allows users to generate and store their private keys while reducing the risk of hackers stealing their digital cryptocurrencies. Furthermore, in the virtual Raspberry Pi environment, RaspiBlitz synchronous data backup is supported by customer firewalls used as an additional security deterrent layer.

Since RaspiBlitz is open-source, anyone can access its code and audit it for security vulnerabilities since it can be publicly viewed online. Additionally, RaspiBitz can be deployed with a preconfigured Operating System (OS) operating system and a Bitcoin node. This capability allows the user to run the node on the Blockchain environment entirely. Similarly, virtual Raspberry Pi contains highly customized features to which users could add more functionalities to the program. These features include but are not limited to running a lightning network as a capability while providing end-to-end multiple blockchain nodes concurrently [10,11].

Since 2012 Raspberry Pi computers have been available through the Raspberry Pi Foundation [12]. Raspberry Pi-s are credit-card-sized, or smaller, computers with under 1-inch clearance for chips and connectors. Different RPi models have varying costs, sizes, speeds, and storage. RPi-s run Advanced RISC Machine (ARM) processors. RISC-Reduced Instruction Set Computers' instruction set architectures is suitable for specialized applications. Such processors focus on smaller numbers of basic machine instructions. These uncomplicated, straightforward instructions lead to less power consumption and lower heat generation. Minimizing power and heat is attractive for IoT devices. Reduced capacity for physical machines may translate to less energy for emulation. Minimizing emulation complexity is valuable in building and maintaining virtual machines. RPi-s use MicroSD and Secure Digital (SD) cards for files and operating systems. Miniature physical versions of these tiny computers support 1 GB or less RAM. More recent versions of RPi-s support up to 8 GB of RAM. Raspberry Pi-s can run a plethora of Operating Systems (OSs). The security of virtual RPi-s is heavily OS-dependent. Both physical and virtual Raspberry Pi-s can be vulnerable to physical/cyber-attacks.

A virtual RPi's host computer can be vulnerable to physical attacks. Yet, virtual RPi-s can be virtualized in large cloud systems and protected from physical attacks. Extending the likelihood of a monitored system can provide security protection against intruders. Next, we discuss the virtual machine Quick EMUlator (QEMU) and the VM manager libvirt. This chapter does not discuss the security of QEMU or libvirt. Instead, we focus on leveraging QEMU and libvirt to host virtual RPi-s that can provide protection [9,13].

The smallest physical Raspberry Pi is the pico—about the size of four cigarettes. It runs a dual-core 32-bit Cortex ARM processor with around 100 MHz and 264 kB RAM. In addition, it comes with a 2.4 GHz 802.11n Wi-Fi device. The largest Raspberry Pi is the RPi 400. The RPi 4B runs a 64-bit ARM processor with 8 GB RAM and a dual-band 802.11an/c with up to 5 GHz. Due to a common architecture, yet in some cases different processors, these machines are similarly capable of being emulated virtually. As these machines are simulated, some virtualized systems can offer emulated devices extra memory and resources while providing an additional security capability layer to similar systems [8,9,13]. Subsequently the advent of automated systems technology, we continued transitioning from manual to automated processes. Modern automation includes sensors, embedded systems, trustworthy computing, Fifth Generation (5G) broadband connectivity, Artificial Intelligence (AI), and Machine Learning (ML). Raspberry Pi-s can be used to enable some of this automation. Raspberry Pi-s can even be embedded in systems to enhance automation. To a large degree, Raspberry Pi-s are becoming key hardware components of the infrastructure. They should be secured to be reliable machines within the infrastructure. Raspberry Pi-s need additional equipment to work with 5G. Fifth-generation wireless adds value by having more and faster connectivity [14,15].

Raspberry Pi-s run billions of operations per second. At the same time, 5G offers 20 GB per second peaks through physical RPi-s, only achieving around 1 GB per second USB3 ports. This speed complements the computational speed of Raspberry Pi-s, where the Fourth Generation (4G) runs at a max speed of about 100+ million bits per second. These technological advances allow the application of Raspberry Pi-s to IoT applications and systems. Virtual Raspberry Pi-s connect through their hosts, which may themselves have 4G or 5G connectors. A host machine may support other high-speed connections. Sometimes the RPi environment directly connects to the cloud network, allowing for substantial data processing and provisioning [8,9,11].

3. Securing virtual raspberry Pi devices

Raspberry Pi-s are miniature computers that can be connected to networks and peripheral devices. RPi-s can work in isolation; hence, remote machines may help enhance security. It is possible to have several Raspberry Pi-s set up to perform joint computations. These collective computations can add a great deal of value. A joint computation may open security issues on physical or virtual networks. Without specialized security knowledge, individuals may deploy Raspberry Pi-s deprived of proper security. Virtual Raspberry Pi-s are less expensive to deploy than hardware RPi-s.

On the contrary, physical RPi-s can be attacked or surreptitiously monitored [9,13]. There is a trade-off exists between having individuals deploy physical Raspberry Pi-s. The virtually deployed Raspberry Pi-s can be easily validated periodically. In extreme cases, these validations are done using the fingerprints of the virtual systems. These fingerprint-based validations are probabilistic. The probability of two fingerprints being the same is exceedingly tiny. In contrast, the probability of an incorrect validation is exceptionally minor. These fingerprints are computed by getting hashes of images from these virtual devices.

Similarly, virtual RPi-s can have standard security configurations. These standard security configurations can easily be fingerprinted, shared, and updated since these systems are virtualized. Many hobbyists set up their RPi-s and their security. If a hobbyist wants to leverage RPi-s for protection, they may be best served to have access to a virtual RPi with security parameters set by a security expert. These physical setups include the installation of the operating systems and configuration. Many operating systems can be deployed on Raspberry Pi-s. The most common one is Raspberry Pi OS. Raspberry Pi OS is a flavor of Debian Linux. It is paired down to work on constrained Raspberry Pi-s. Ubuntu desktop images are 3 GB, but Raspberry Pi-s operating system images range from ½ GB (Raspberry Pi OS Lite) to closer to 1 GB. Linux systems such as Ubuntu can support giant amounts of persistent storage. RPi-s have persistent storage that is 64 GB, 16 GB, or even 8 GB or less [9,13].

Security best practices may include small portable devices such as virtual RPi-s. These virtual devices can act as independent security systems. Different security tools can be on board other virtual systems. In the same way, it is common for individuals to ignore checksum validations of fingerprints while installing software or systems. Virtually installed Raspberry Pi-s can automatically include this critical step in the setup process. These devices can even be validated on a blockchain. Pi-s can store the fingerprints for each version to ensure the data input to the fingerprint functions has not changed. To provide additional layers of security, Ferret Pi may be suitable. Ferret Pi offers protection to Raspberry Pi servers. These servers are deployed on virtual Raspberry Pi-s. Ferret Pi is a File Transfer Protocol server with secondary attributes. Such attributes include self-destruction if the persistent storage SD card is removed for over five minutes. This persistent storage destruction can be done by formatting the SD card when remounted after five minutes of removal. Ferret Pi can be deployed along with Raspberry Pi as a Secure File Transfer Protocol Server. Once attackers understand how Ferret Pi works, they can take steps to avoid its storage destruction [13].

Wi-Fi, Bluetooth, 4G, and 5G allow RPi-s to connect with the world. This process can be carried out by monitoring or managing data. Long Range Wide Area Networks (LoRaWAN) may be deployed to help RPi-s work and act as IoT gateways. Virtual RPi-s can use these communication systems through their hosts. LoRaWAN transmission is based on a star network topology. Each node connects to the LoRaWAN gateways. These gateways connect to the centralized network server environment. LoRaWAN is an open-source technology with good coverage and low power consumption [8,9]. LoRaWAN's drawbacks include but are not limited to, a lack of network infrastructure, roaming agreements, and extreme messaging capabilities limitations. LoRaWAN Communications take much work to jam—these solutions can be deployed with virtual Raspberry Pi-s that use Physical Communication Dongles (PCD). PCDs are small devices plugged into computers and/or servers. These devices are deployed to facilitate certain software enablement. Dongles can be used as small adapters that enable specific software. These small devices can similarly be deployed as security measures allowing the use of particular software [8,13].

Once a computer virus can be isolated, we can hash its machine code into a fingerprint. Each virus has a fingerprint that uniquely identifies it, with high probability. SHA256 always outputs 256-bit fingerprints. There are no known inputs that have the same SHA256 fingerprints. Since there are at least several million computer viruses, they require several million digital signatures to identify each virus uniquely and quickly. One signature, based on SHA256, requires $32 = \frac{256}{8}$ bytes. One million signatures require 32 million bytes of storage [9,13]. Network traffic can be tracked using the Wireshark program. Wireshark may need ½ GB of RAM and ½ GB of persistent storage. These long-term storage files can be analyzed, requiring more RAM and storage. Data mining for these files can be challenging due to physical Raspberry Pi limitations [13].

Raspberry Pi-s can be deployed as an integrated security and surveillance system. Raspberry Pi-s functions span processes, techniques, applications, security policy, implementation, and deployment practices. Raspberry Pi-s can be a proxy or bastion host for a network. A bastion host checks credentials and allows appropriate users or programs to pass through a secure network. Bastion hosts can be virtual when deployed to the cloud. Computer proxies perform operations in place of other machines. Reverse proxies send out data in place of different devices. Proxies and reverse proxies can save information about their interactions outside their networks. These logs are helpful for security analysis, and proxies or reverse proxies can be made from simple processes. Techniques used on virtual machines vs. physical devices are the same. Having virtual RPi-s act as proxies or reverse proxies is natural. If the Raspberry Pi-s security is breached, the entire system's security can be questionable [8,13,16].

4. Raspberry Pi automation for advanced IoT systems

Virtual Raspberry Pi-s are susceptible to vulnerabilities, threats, and risks. These user concerns range from denial of service attacks, unsolicited user access data breaches, and malware infections. As a result, the industry must develop virtualization technology solutions to minimize and prevent emerging risks, vulnerabilities, and threats against the hypervisor layer. Some of these concerns can be mitigated by implementing new strong security measures. More recent antivirus software, access controls, and firewall capabilities may prevent these security issues; continuous OS and patch virtualization software update is critical to enhanced system security. Businesses and organizations using virtual Raspberry Pi must train their users. Lacking such security measures may result in human error and social engineering attacks. These errors can also threaten virtual Raspberry Pi instances [2].

Virtual Raspberry scalability and reliability are based on the virtualization software and hardware of the run system. The creation of virtual Raspberry Pi instances can be simultaneously carried on, meeting the system scalability and capacity terms. Besides, these instances can be cloned and deployed simultaneously. This process allows for smooth system scalability and reliability. Virtualization is an integral solution for performance overhead despite any limitations the system may incur. This process can be achieved based on the specific cybersecurity application. The reliability process focuses on virtualization software and hardware solutions. Virtualization can pose various security risks that may lead to system crashes. Users may experience system and performance issues impacting the reliability process if this occurs. Regular backups and system updates can quickly mitigate this issue. Such measures can help reduce risks and guarantee that the virtual Raspberry Pi environment is more reliable [10,17]. High-tech advances allow for the intersection and interaction of Raspberry Pi and IoT systems. Raspberry Pi-s are helping grow new economic systems [9,13]. These tiny devices can add computational power in many places.

Raspberry Pi-s add capabilities that can be deployed to help set up low-cost servers or services. Object-oriented programming has several crucial attributes, including objects encapsulating methods or functions and object data. These methods have access and may mutate an object's data. In data-oriented programming, immutable and lightly structured information flows through a system as it is processed. Other paradigms might evolve using IoT devices and small computers like Raspberry Pi-s. Complex tasks in specific environments may require several context-sensitive services. In such cases, coupling exact domains with several computational services may be helpful. For example, in tracking face recognition using a drone, securely transmitting a stream of recognized faces with their locations while validating the system requires several services [10]. In addition, the system should regularly validate its state and security. At least one service with the entire system validates an encapsulation of the other services and the whole system. Our interest is in low-resource systems such as Raspberry Pi-s. Deploying these devices virtually can be problematic; virtual RPi-s are not containers. Containers share underlying infrastructure and resources. The main objective in this context is to have a good deal of independence and self-reliance for the RPi-s. Due to shared resources, containers are less expensive than virtualized machines. Consolidating a virtual RPi on another device may limit the essential hardware functions. Adding a virtual machine could require more capacity and open up other vulnerabilities [9].

Today there are many small or constrained computers in use. These examples include ODB2 car interfaces, smart-home controllers, parking meter controllers, and even some sneakers with computers. In the case of ODB2 interfaces, there have long been small portable computers to query these interfaces. Smart homes have security systems operated by local sensors and cameras, and their data can be relayed to the Internet and the cloud. Monitoring smart homes requires local controllers. Raspberry Pis can be used as controllers or Internet Gateways. Virtual machines must be secure when deploying IoT solutions while considering the costs, time, technology, assets, and resources [8,13]. Virtual Raspberry Pi-s may be scaled vertically and horizontally. Vertical scaling is to add more resources to a virtual machine. Horizontal scaling is to add more virtual machines to work on the same challenge. Such virtual RPi's provide additional possibilities for data-driven applications. RPi-s can be coupled with data in a technological environment. Reproducing specialized results requires the same data inputs, the same algorithms, and sometimes similar environmental factors. Large horizontally scaled instances of virtual computers, such as Raspberry Pi-s, can harden security through periodic updates, patches, and reconfigurations. However, these updates may lead to security issues [9,13].

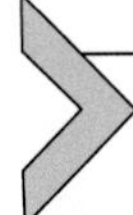

5. Virtualization for raspberry Pi and the internet of things

In 1972 IBM unveiled its first virtual machine, practically known as 370. Virtualization allows users to share data, systems, and other resources and information. The process enables the creation of systems at a level of abstraction. The concept of virtualization spans virtual computer hardware systems such as network resources and storage devices. It involves memory virtualization, virtual functions, and language virtual machines, namely (.Net and Java Virtual Machine) [18,19]. Virtualization refers to emulating entire computers.

There are several types of virtualization though our focus in this chapter is on the guest machine emulation using software to mimic the guest machine's behavior broadly. Virtualization provides alternate copies of devices—these virtualized computers are Virtual Machines (VMs). In a Virtual System (VS), guest machines are emulated by a host machine. In RPi-s, we are interested in full software emulation. In VS, physical computer users can run several Operating Systems (OS). These OSs can be run on a single host system. Hence, the guest machines run in the VMs [18,20].

To virtualize an entire computer, emulators are managed by hypervisors. When emulating in the virtualized environment, physical machines may host multiple virtual machines [9,13]. Live virtualized systems may be migrated to other hosts. These alternate hosts may be in different locations through virtualized systems that can be saved and copied. Many teams may share and examine these guest images and their states simultaneously. This process can be easy to execute using virtual machines to add value for securing systems equipped with virtual RPi-s [8,9].

Virtualized systems like Raspberry Pi-s may be quickly tested in many configurations. This process can be compared to reconfiguring hardware manually. Two types of Virtual Machine (VM) replication are real-time and Point-in-Time (PIT). Real-time replication is a process that allows data to be digitally imaged or copied to a replicated machine anytime. Such a process may occur when security events are detected or anticipated. PIT replications are usually specified at specific times for routine backups [21].

Systems backups occur when VMs create/clone another VM version(s) that can be located on the host system. Periodic PIT replications can be studied to understand security better. Computing fingerprints and their states can preserve these replication image processes, while the techniques may focus on improving system security. These techniques can help understand the systems, improving reliability, performance, high availability, and uptime [21].

Virtualized IoT devices can be done by migrating data and applications from one host to another. The host machine may be used to understand computation in specific environments. Next are examples of VMs attributes and how they help with security [18–21]:

- **Isolation:** It is a logical or physical decoupling or separation of VMs. It can prevent a failed or compromised system from affecting other machines. Although the decision to eliminate or isolate the system from communicating with other peer devices in the virtualized environment, limiting a machine from interacting with other VMs may be vital to the system's continued security or proper performance. When VMs are isolated, they cannot directly share their processors, memories, and other resources. Virtual guest machines access data from their host and share resources with other guest machines. A VM's host may prevent its guest from communicating with other systems since it may control access to and from a guest.

- **Virtual Machine Migration**: A guest VM may be migrated from one host to another host. The process involves moving allocated resources from one host computer to another host. For example, a host machine can migrate data, configurations, and virtual hardware, to other host systems. If a VM is compromised or vulnerable to attacks, migrating resources to a safer host machine may minimize vulnerabilities.
- **Replication**: Replication is a method for backing up VM systems. Replication may be used to migrate VM images. If computer data is corrupted, lost, or compromised, backups are used for restoring VMs. Backups can be used to validate systems, data, and copies of VMs can be easily replicated. These backup copies allow independent and distributed security or forensic investigations.
- **Cloning**: Making a full duplicate version of a VM is done through a cloning process. Horizontal scaling uses cloning. Cloned machines may have their Media Access Control addresses (MAC addresses) and 128-bit value Universal Unique Identifiers (Globally Unique Identifiers) set to any value. Clones may copy and change unique ids. This approach is often regarded as a security vulnerability. Compromised systems can be cloned and monitored to understand an attack focus better.
- **Ease-of-Provisioning/Virtual Machine Provisioning**: The configuration of VMs is done using scripts. These scripts can be automatically modified and updated. The process is helpful for virtual machine security. Different configurations offer extra protection. Some systems operate best with other resources, that is, large amounts of memory. Virtual machines make it easy to add memory. The process leads to testing the security of unconfigured devices, fully configured systems, and readily installed OSs.
- **Agility**: Agility involves how systems may respond to computer configuration, performance, and challenges that could impact their operations. The VM's redundancy is based on interconnected systems' ability to provide an agile response to anomalies and challenges. This can help secure systems, including RPi-s. The easier it is to update a system in response to an issue, the better for security.

Several complex instructions set architecture processors have virtualization features built in. As examples, Intel Virtualization Technology/Intel VT-x and Advanced Micro Dynamics Virtual Machines (AMD-VMs)/AMD-Vs architectures have special instructions for emulating guest machines. These guest modes are entered and exited through particular instructions. They prevent guest machines from accessing privileged instructions on the host system. Both VT-x and AMD-V architectures are complex instruction set architectures. Due to their commonality and specialized virtual instructions, VT-x or AMD-V systems commonly host virtual Raspberry Pi-s [18,20].

5.1 Hardware

Virtual Machine Managers (VMMs) are VM guest managers and monitors. VMMs manage virtual machines for a host computer. VMMs can run on the host machine though some may run on the bare metal of a host. Whereas other VMMs run on the operating system of a host. VMM can run on a virtual guest's operating system. We do not cover the security of VMMs in the chapter. Virtual Raspberry Pi-s can be security gatekeepers, work with digital forensics, and contribute to or maintain public ledgers of IoT data. A very well-explained view of networks is in Howser. Howser illustrates high-level network concepts on physical Raspberry Pi-s. This concept remains vendor agnostic for network equipment. Unfortunately, security on RPi-s is not a central focus of Howser's book [8,22].

6. Blockchain methods and pigeonhole principle

In 2008 Satoshi Nakamoto proposed the first blockchain. Nakamoto's decision was illustrated in a whitepaper entitled "Bitcoin: a Peer-to-Peer Electronic Cash System." Satoshi created the first Genesis block in the same year as part of his blockchain technology innovation. Integrating blockchain functionality into systems significantly influences innovation. Nakamoto's identity remains anonymous despite his innovative blockchain concepts and technological approaches. When Satoshi was created, Bitcoin and blockchain technologies used a nom de plume to communicate despite his disappearance in 2011. Blockchains are a set of records that consist of daisy-chained blocks. Each block is created using a probabilistic consensus mechanism. This probabilistic consensus leads to the time between blocks being probabilistic for some blockchains [23–25].

Blockchains are public or private ledgers, payment systems, and cryptocurrencies. They are also dispersed public ledgers that record Bitcoin transactions. In this context, a ledger lists transferred accounts and items or values. In addition, ledgers can log security events and system status generally maintained by several computer networks deployed worldwide. Blockchain ledgers are made of blocks. Each block is linked to another using a secure cryptographic method. In the blockchain, all transactions are subject to some verification or validation processes. Instead of using participants network-based approach to verify participants, in a blockchain, there is no need for centralized authority for continuous verification and validation. Validating blocks and adding values to blocks is inexpensive. In general, Raspberry Pi-s can add data to blocks and validate blocks. In a blockchain, every block has a designated cryptographic hash supplied by a previously assigned block. This method is akin to a timestamp, in which each hash chain element carries verifiable transactions in each block. These timestamps can be probabilistic though the required period between timestamps may be based on an expected time difference [23].

Virtual Raspberry Pi is a suitable platform that can be used to develop and test blockchain, cybersecurity, AI, and ML applications. Virtual Raspberry Pi allows multiple virtual hosts to run parallel when running in a single physical machine. Such ability also allows for platform scalability and efficiency. The virtual Raspberry Pi can be deployed to create and test neural networks in AI and ML environments. It can also run simulations and build ML models. In the cyber environments, it can run penetration testing toolkits, security solutions creation, and vulnerability assessments of designated environments.

A virtual Raspberry Pi can provide consensus algorithms, build dApp, simulate smart contract functionality, and test. Given the rapid technological shift, developers and researchers might immensely benefit from virtual Raspberry Pi capability and offerings. Virtual Raspberry Pi usage, scalability, and adaptability can be integral tools in similar technologies, such as Natural Language Processing (NLP) and cybersecurity applications. It can be used in generative neural networks and NLP for algorithm development and testing. Such activities may include text summarization, language translation, and speech recognition capabilities. As an evolving technology, virtual Raspberry Pi can be a trusted platform for building algorithmic models to generate text, images, and other data types. In addition, educators, developers, and researchers can use virtual Raspberry Pi to explore, innovate, and discover new technologies. A blockchain is a vital tool that users and organizations need to boost security, prevent theft of private keys, and minimize risk to digital cryptocurrencies. Virtual Raspberry Pi is critical in blockchain technology [26,27].

Virtual Raspberry Pi-s with blockchains can be used for cybersecurity applications. Integrating virtual Raspberry Pi and blockchain technology gives developers the security and development of decentralized and distributed network solutions. This capability allows virtual Raspberry Pi and Blockchain to share and store sensitive data for cybersecurity applications. Virtual Raspberry Pi is used in a controlled environment to test several cybersecurity applications. This method is typically carried out before these cybersecurity applications are deployed to identify and mitigate vulnerabilities. Blockchain and virtual Raspberry Pi can enhance cybersecurity practices while helping minimize security risks, threats, and vulnerabilities. Small virtual RPi-s networks can implement blockchains validated on larger public blockchains. These networks of RPi-s are building sidechains. Two types of blockchains are proof-of-work and proof-of-stake blockchains [17,28,29].

6.1 Proof of work

Proof of work blockchains requires certain levels of work to participate in building the blockchains. Miners create and verify blocks in the blockchains. Their distributed ledgers are usually public, though on average, miners must work hard to throttle them to the limits of their computational commitment. Almost any computer can participate in mining. Miners are limited by their power even though this approach makes proof-of-work blockchains consume much power. This egalitarian method is mainly embraced by people who have concerns about monopolies.

6.2 Proof of stake blockchains

In proof-of-stake blockchains, build blockchains limited to participants that can demonstrate a stake in the system. These limitations give a sense of security to the system. The distributed ledgers may be private for proof-of-stake blockchains. Similarly, such participation limitation allows monopolist power. Proof-of-stake blockchains may consume less energy than proof-of-work blockchains.

6.3 Blockchains to enhance virtual RPi security

Distributed ledgers can store security events. For example, sensors detecting environmental changes aim to be irrevocably recorded in blockchain ledgers. Keeping security events allows later analysis and may even maintain evidence in the case of alleged crimes. Similarly, using sidechains made by virtual RPi-s keeps most of these records off large and expensive blockchains. Sidechains may use proof-of-stake, saving resources and allowing local storage and validation. Ideally, each virtual RPi can be cloned and quickly provisioned to run blockchain nodes. In the virtual case, individual blockchain miners can be horizontally or vertically scaled and migrated to new locations while continuing to generate their blockchains.

6.4 Pigeonhole principle

Placing n pigeons in k pigeonholes requires that there must be at least one pigeonhole containing at least ceiling (n/k) pigeons. Ideally, good deterministic message digest hash functions appear to distribute their outputs uniformly over their range. Given a suitably extensive range, such ideal deterministic message digest hash functions give output fingerprints unique for all practical purposes for different inputs. However, the pigeonhole principle requires output collisions when the number of inputs exceeds the possible outputs. For example, consider having 2^{256} possible pigeonholes. Having $2^{256} + 1$ separate inputs is impossible to list since it is impossible to list $2^{256} + 1$ numbers even with a computer. The likelihood of two different inputs having the same output using a good message digest hash function is very improbable. The below image illustrates our recommended three-hash digest and hashes chain approach [30,31] (Fig. 1).

Digital chains of custody are timed hash chains used to secure digital evidence. The concept can be applied to digital forensics. It can be implemented on virtual Raspberry Pi-s. Digital custody chains are digital transactions typically written to record the custody sequence. This method includes but is not limited to controlling, transferring, and analyzing physical dispositions such as electronic evidence. These digital chains involve a logical sequence that records and tracks evidence for potential digital devices or crimes [12].

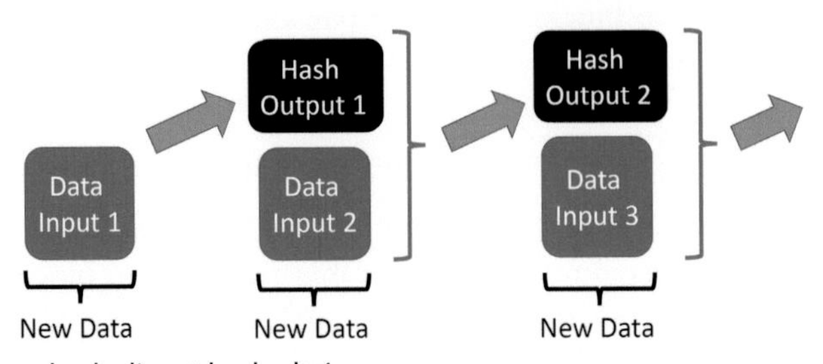

Fig. 1 A three-hash digest hash chain.

6.5 Message-digest and cryptographic hash algorithms

A hash chain is a sequence of hashes—each hash function application's input includes the previous hash function application output. In particular, SHA-256 has 256-bit results. Hash chains using SHA-256 and SHA-512 require at least $2^{256} + 1$ and $2^{512} + 1$ inputs to guarantee a hash collision by the pigeonhole principle. If we applied either of these functions to hash RPi images, we would need an infeasible number of RPi-s to guarantee two hashed images are the same so long as all these images are unique [30,31]. The following are several versions of SHA in use to date: "SHA-2, SHA-224, SHA-256, SHA-384, and SHA-512" [31,32].

Government and academic researchers developed Cryptographic Hash Algorithms. The National Institute of Standards and Technology (NIST) developed SHA. SHA aims to "generate fixed length digital fingerprints, hashes of input data." These solutions are used for data integrity and authenticity verification.

In 1992 the National Institute of Standards and Technology tested and approved the SHA, and the Standard was announced in 1993. The Digital Signature Standard, part of the Secure Hash Standard, developed the product [33].

The SHA hash functions were developed following the MD4 algorithm. These hash algorithms are deterministic and inexpensive to run. Raspberry Pi-s can run these hash functions to validate proof-of-work and proof-of-stake hash chains. These capabilities may provide IoT platforms with the secure and transparent solutions they need as an additional layer of protection. Raspberry Pi-s support a range of versions of secure hash algorithms solutions. Message digest hash functions can compress large amounts of data into tiny digital images [30–32]. Given some notable advances in the cryptographic community, Raspberry Pi can quickly implement blockchain structure and hash techniques solutions. These capabilities may provide IoT platforms with the secure and transparent solutions they need as an additional layer of protection. Some IoT platforms are mini-systems built with standard message digest hash functions and public key systems. Raspberry Pi supports a range of versions of secure hash algorithms solutions. Message digest hash functions can compress large amounts of data into tiny digital images [31,32,34].

Digital chains of custody are timed hash chains used to secure digital evidence. This method can be applied to digital forensics. It can be implemented on virtual Raspberry Pi-s. Digital custody chains are digital transactions typically written to record the custody sequence. This process includes but is not limited to controlling, transferring, and analyzing physical dispositions such as electronic evidence. These digital chains involve a logical sequence that records and tracks evidence for potential digital devices or crimes [5].

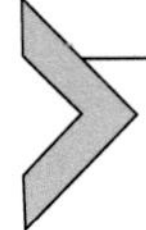

7. IoT devices with smart surveillance monitoring systems

IoT device security through innovative surveillance monitoring protocols is an area that continues to evolve. A machine-to-Machine surveillance approach motivates developing automated systems to provide continuance surveillance to residential and business infrastructures. Surveillance systems can be deployed singly or collectively via edge computing in an IoT environment. A thoroughly integrated approach to fog computing solutions can support cloud operations [35]. Organizations may invest in advanced automated surveillance systems that require limited to no human interaction. A virtual RPi may migrate to follow a person while building a hash chain or blockchain carrying proof for the person's travels, face recognition, and event logging. Documenting the time and date an event occurs is essential to surveillance. The intent is to answer, "who did what when?" [35]. These events can be physical events as well as digital events. Some examples of this process include a video recording of a person's entry into a location. While many cameras can record dates and times, sometimes these dates and times can be recorded in error or overridden [35]. Videos or images can be edited, and deepfakes can be created. Publicly verifiable timed hash chains, such as blockchains, can allow documentation of the approximate time and dates of events. Given the correct structures, timed hash chains can be converted to digital chains of custody [19]. These digital chains of custody may be enhanced for recording evidence with approximate dates and times. The process allows virtualized Raspberry Pi-s as virtual devices can be migrated to follow a suspect. Deepfakes are more troublesome since they can fake events that never occurred. Many validated locations for people can disprove deepfakes. Suppose an individual is validated in other places when a serious fraud suggests they are somewhere else. In that case, the deepfake can be discounted quickly. This process is due to the virtual RPi-s ability to place fingerprints of images, face recognition, and movements in a secure blockchain [35].

8. Virtual raspberry Pi security

Virtual Raspberry Pis are secure and trusted virtual environments. Virtual machine software can be used to create these secure virtual environments. These environments usually are designed to run on various computer systems. Virtual Raspberry Pi can be used in cybersecurity applications to create a safe and remote platform environment. It can be used for network security-related techniques and tools experimentation. This method focuses on conducting penetration testing and exploring vulnerabilities. The concept has been applied to developing defensive strategies to combat cyber-attacks and threats. Most organizations use a secure platform environment as a security sandbox to assess threats and cyber-attacks that can be launched against virtual and physical Raspberry Pi. This security solution is cost-effective and can be used to educate users on how to detect, mitigate and neutralize known and emerging threats against virtual or physical Raspberry Pi-s. VMWare and VirtualBox virtualization systems are integral in creating cybersecurity applications. This virtualization software monitors, scan, analyze the network, and test malware in the system [29].

Users with limited access to physical Raspberry Pi can use this tool. Some users interested in experimentation or penetration testing may opt for other cybersecurity tools available on the market [29,36]. Hence, a virtual Raspberry Pi is a replicated version of the physical Raspberry Pi machine. This simulated version can be launched on a host machine. Virtual Raspberry Pi's advantages and disadvantages include but are not limited to [37,38]:

Advantages

- **Cost-Effectiveness:** Virtual Raspberry Pi can be used free of charge; physical Pi-s must be purchased. Users must acquire hardware components and other peripheral devices that could be costly.
- **Flexibility:** Virtual Raspberry Pi-s can be installed and quickly configured without adding physical accessories and components. Users can customize a virtual Raspberry Pi to perform detailed requirements and functions without involving hardware changes.
- **Convenience:** Users can access virtual Raspberry Pi resources using their host machine anywhere. To complete such a process, users must have access to the host machine, which allows them to conduct penetration testing and development remotely.
- **Safety:** To conduct safety experimentation without risking damaging physical Raspberry Pi components, users must apply safety when using virtual Raspberry Pi-s.
- **Multiple Instances:** This option allows virtual Raspberry Pi to launch multiple instances within a VMM. This process allows launching such activities on a single host machine to improve testing and experimentation functions.

Disadvantages

- **Performance**: There is a performance difference between virtual and physical Raspberry Pi. The difference comprises tasks involving high processing power. In addition, there can be conflicts between the graphics drivers of the VMs and the host machine.
- **Hardware Access Limitation**: Virtual Raspberry Pi might access General Purpose Input Out (GPIO) and camera modules. Instead, physical Raspberry Pi can fully access GPIO and camera module hardware components.
- **Lack of Physical Experience**: Virtual Raspberry Pi does not have comparable experience with physical Raspberry Pi. This experience is integral to the development and learning purposes.
- **Compatibility Issues**: Several software and hardware components do not work well with virtual Raspberry Pi. Lacking system compatibility problems between host machine OS systems with virtual Raspberry Pi is one of the issues that both Raspberry Pi-s continue to face.

The security of virtual Pi is central to overall operational functioning. This approach applies to users running the physical Raspberry Pi hardware components. The list below illustrates security measures that must be incorporated to guarantee the virtual Raspberry Pi's safekeeping and protection. The security and safety measures are [39,40]:

- **Secure Host Computer**: The host machine and virtual Raspberry Pi have equal security requirements. To guarantee proper security, users must ensure that host computers are malware-free. This security measure must ensure that the host is malware-free and that security patches are updated periodically.
- **Secure Network**: Virtual Raspberry Pi run on a secure network. Encrypted solutions and firewalls are fundamental in protecting virtual Raspberry Pi. This process allows for the prevention of unauthorized access to the secure network. Users must disable any network services while limiting network access involving trusted devices.
- **Use of Strong Passwords**: Virtual Raspberry Pi user accounts must use robust passwords and defaulting usernames. Strong passwords can prevent unsolicited users from accessing the environment.
- **SSH Security Features Enablement**: A secure shell (SSH) is critical to virtual Raspberry remote access. The configuration of SSH is vital to the functioning of a virtual Raspberry Pi. Configuring SSH, strong encryption, and disabling the root login feature guarantee more security for the

virtual Raspberry Pi. Users must rely on a public key authentication feature to log into a virtual Raspberry Pi securely.

- **Software & Operating System Continuous Updates**: Regular software and OS updates are fundamental for the healthier functioning of virtual Raspberry Pi platforms. Keeping a secure system while patching security vulnerabilities offers virtual Raspberry Pi the level of security these platforms need to be risk and malware-free.
- **Data Backup**: Any data stored on the virtual Raspberry Pi must frequently be backed-up. These regular backups help prevent any losses should there be any data breaches.
- **Disabling Unnecessary Services**: Disabling virtual Raspberry Pi unnecessary applications or services is essential to better platform functioning. This process helps reduce attack surface while improving system security. Implementing these enhanced security measures makes Raspberry Pi less vulnerable to known and unknown attacks against the virtual Raspberry Pi.

Virtual Pi comes in many flavors and options, some of which include [41–45]:

- **Raspberry Pi Desktop**: An open-source software program simulating a computer's Raspberry Pi desktop environment. In a Linux-based graphical user interface (GUI) environment, the Pi Desktop provides a user-friendly and familiar interface. This interface supports the OS—Operating System. Users can perform tasks like file management, application running, web browsing, and other related tasks. As software, it can be installed on a Raspberry Pi. Users can customize Raspberry Pi Desktop despite its options for customization and altering the appearance, adding applications, and modifying settings to satisfy the user's requirements.
- **Raspberry Pi Emulators**: As software programs, users can use emulators to simulate Raspberry Pi hardware. Emulators can be used to test and learn how to use the system without procuring a physical Pi machine. A list of Raspberry Pi emulators comprises—QEMU, VirtualBox, and Raspberry Pi Emulating processes using a Docker environment. Most emulators can provide popular Raspberry Pi OS pre-built imagery. This approach generally focuses on OS such as Raspbian that can be set up and deployed faster. Several emulators can simulate the Raspberry Pi hardware. Raspberry Pi Emulators allow users to run programs as they would use the actual Pi.

- **Raspberry Pi Cloud Services**: AWS and Azure can offer VMs that allow users to access Raspberry Pi in several enterprise server environments. Additionally, these cloud-based platforms can deploy VMs with Raspberry Pi capabilities. As a result, users can remotely create and manage multiple environments of Raspberry Pi. Such a process usually takes place without involving physical Raspberry Pi. Popular Raspberry Pi services span AWS, Google, and Azure Cloud. Developers and hobbyists are seeking testing environments as well as deploying assorted projects. This process can be employed without purchasing hardware or infrastructure management solutions. Users may purchase subscriptions or make monthly/annual payments to access some of these services.
- **Raspberry Pi Virtualization**: Virtualization software such as VirtualBox and VMWare. This software creates virtual Raspberry Pi environments on one or multiple computers. This process focuses on creating a virtual Raspberry Pi computer using VirtualBox and VMware. Users can test applications and create development environments in the Raspberry Pi virtualization setting. This concept also focuses on running a Raspberry Pi environment through a single computer in multiple instances. In cloud computing, virtualization can create and manage virtual Raspberry Pi via remote server environments.
- **Raspberry Pi Remote Access:** Users can access Raspberry Pi virtually. This process is achieved using SSH, Remote Desktop Software, or VNC—Virtual Network Computing (VNC). VNC can be deployed to control and allow remote access to Raspberry Pi's desktop environment. In a Raspberry Pi environment, users can remotely access projects. This method makes it much easier for users to debug, update, and monitor the system. In the Pi setting, users can remotely access resources. This process can be accomplished by ensuring the device has excellent network connectivity and attaching the keyboard to the physical Raspberry Pi. The process also requires proper remote access software user configuration on Raspberry Pi and the remote device. SSH can be installed to give users access to the Raspberry Pi terminal environment. In the Raspberry Pi environment, SSH can access the Raspberry Pi command line interface virtually and execute the command securely.

Raspberry Pi-s provide advanced security capabilities. As intelligent surveillance systems, virtual Raspberry Pi-s can be deployed virtually anywhere. Virtual RPi-s can be implemented as intelligent surveillance sensors. These motion sensors may be used to run applications by migrating the virtual machines to monitor events. For years, sensors have been used to measure temperature, humidity, and motion detection. These are low-cost computations and do not need an entire computer to work. Virtual Raspberry Pi systems can provide advanced and robust surveillance. RPi-s play a pivotal role in machine learning and AI. It applies/identifies face recognition, motion tracking, event correlation, and even causal event evaluation. Maintaining alternate systems, such as RPi-s, comes with variable costs too. Virtual RPi-s can be deployed and monitored on a large scale. Rpi-s can be replicated virtually or horizontally as needed. Continued video surveillance technology advances and causal models can be easily integrated into virtual RPi-s. These integrations can be done with remote virtual systems such as Raspberry Pi-s to build improved sensory devices continually [35].

9. Raspberry Pi for QEMU machine emulator/virtualizer

QEMU is a popular and flexible set of virtual machines. QEMU virtual machines can be used to emulate the Raspberry Pi-s. QEMU emulates ARMs that act as CPUs for Raspberry Pi solutions. ARMS require support chips on their motherboards for easy function and focus on simulating other support systems (Table 1).

ARM systems are attractive for IoT devices since they use limited power. ARMs have small and straightforward instruction sets. The instruction set architecture of ARMs is small, simple, and regular. Complex instruction set architectures are richer, though their complexity comes with some costs. The most common classes of complex instruction set architectures are ×86 and ×64. Both ×64 and ×86 are more general-purpose processor architectures with comprehensive instruction set architectures. A downside for ARM processors is that they require specialized system boards. These system boards must be emulated. Emulating additional hardware adds complexity and potential vulnerability since these unique boards may come in wide varieties and configurations. Kernel-based Virtual Machine (KVM) is essential in

Table 1 QEMU works with KVM Libvirt.

System	What	Link
QEMU	Processor emulator	https://www.qemu.org/
KVM	Kernel-based virtual machine	https://www.linux-kvm.org/
Libvirt	Virtualization management tool	https://libvirt.org/

virtualizing Intel VT-x and AMD-V instructions and extensions. The ×86 and ×64 extensions manage virtualization on the hardware platforms.

Libvirt allows for the VMs on QEMU management and implementation. KVM enables the emulation of VT-x and AMD-V extensions. Fig. 2 shows two Raspberry Pi-s isolated by different instances of QEMU. Fig. 3 illustrates three Raspberry Pi-s isolated on three separate models of QEMU but a shared libvirt-controlled virtual network. Libvirt and QEMU VMs run on the Ubuntu instance as the host machine. The Ubuntu VMs can run on a VirtualBox VM or similar VMs [46,47].

Despite being isolated, RPi solutions can be interconnected for collaboration and performance RPis communicate through the host computer. This is because the two QEMU VMs on the host are independent.

Despite being isolated, RPi solutions can be interconnected for collaboration and performance RPis communicate through the host computer. This is because the two QEMU VMs on the host are independent.

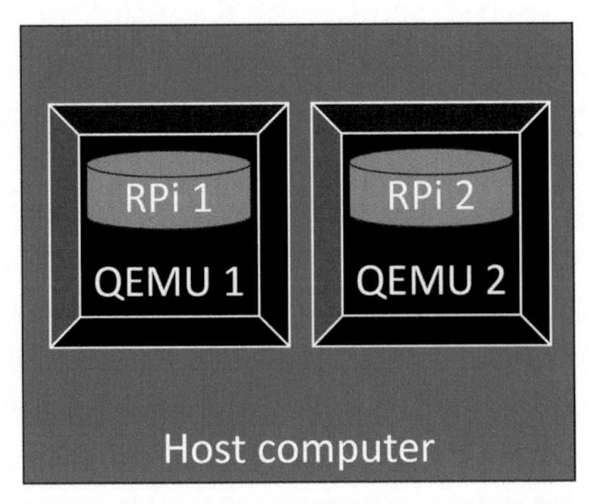

Fig. 2 Two separate instances of QEMU hosting two virtual Raspberry Pi.

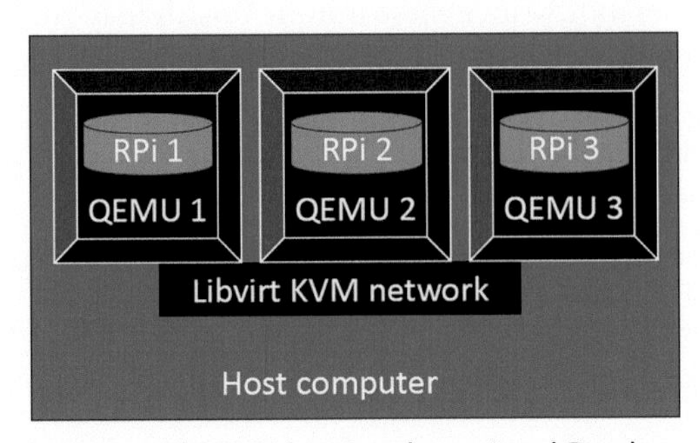

Fig. 3 Two libvirt instances of QEMU hosting three virtual Raspberry Pi.

10. Pi-hole message tracking and raspberry Pi

Pi-hole is a network-wide ad blocker built to protect Raspberry Pi networked devices. Pi-hole is built to replace adblocker software installed on every device and browser. Advertisements can be intrusive; they may follow computer users, exposing them to attacks. Tracking users is a severe security breach; following users can enhance social engineering attacks. Pi-hole is built to block ads trying to access the system. Pi-hole can be installed on VMs or containers to provide seamless services. These services include the process of installation on Raspberry Pi OS. Pi-hole acts as a gatekeeper for Virtual Private Networks to block unwanted ads and track user activities within the network. At the DNS, it prevents Internet denizens from penetrating devices or computers. These ads may contain malware that must be filtered or blocked [48].

11. Network security for raspberry Pi

Small computers such as Raspberry Pi-s can screen threats affecting their environment. Open-source packet analyzers and network security toolsets such as Ettercap and Wireshark can scan network packets for potential threats. Virtual hosts can be created to avert masqueraders' activities against single or multiple systems.

Intrusion Detection Systems can be used alongside Ettercap to protect the network traffic. The tool can gather network activities, ports, and Media Access Control. Implementing Raspberry Pi in a network environment provides advanced security solutions such as cryptographic hash algorithms to IoT devices in the enterprise. Failure to secure connectivity between Raspberry Pi-s and IoT devices could result in device and application data theft or malicious attacks. Given their low complexity and ease of deployment, Raspberry Pi-s may be an alternative solution for honeypot technology.

A virtual RPi that acts as a honeypot can be dynamically isolated from production systems. A normally functioning virtual RPi can dynamically turn into a honeypot while mocking a natural approach. Raspberry Pi-s, Snort IDS, and honeypots can enhance information collection on attackers [28,49].

12. Distributed computation and raspberry Pi-s

Understanding Raspberry Pi's vulnerabilities gives us a necessary critical frame of reference. RPi-s can be exposed to insider and outsider threats.

Security threats include eavesdropping, man-in-the-middle, and signal jamming. Some attacks, such as signal jamming, are on physical Raspberry Pi-s or peripherals. Physical or virtual Raspberry Pi-s can be embedded into other platforms. A root of trust is a trusted place to start security. Cryptographic methods can protect message tracking and data transit. That is data at rest or data in transit [49,50].

Small clusters or networks of RPi-s may be able to fake natural systems effectively. These clusters or networks can be used as honeypots to understand adversaries. Since RPi-s are small and inexpensive, and virtual RPi-s are mobile and agile, these honeypots can be set up quickly. Devices like Raspberry Pi-s have impacted resource deployment in distributed computing. Consequently, applying cryptographic methods to Raspberry Pi Server Clustering (RPSC) may be helpful. RPSC can be deployed to provide file encryption and decryption capability between computers [51].

Open Multi-Processing (OpenMP) and Message-Passing Interface Chameleon (MPICH) libraries are used in parallel computing environments. Additionally, distributed or parallel systems, including OpenMP, hybrid, and MPI—Message Passing Interfaces can be deployed to provide parallel or distributed computation. Securing these distributed computation systems is critical. Interesting security developments here include the pairing concept of identity-based encryption. The basic idea here is that some identifying information is included in the encryption and decryption.

The Tiny Encryption Algorithm (TEA) is a symmetric encryption algorithm for constrained devices. TEA can be implemented for data encryption and as a message digest hash function. Raspberry Pi-s can run standard TLS/SSL solutions. MPI can provide synchronization between communication threads while in transit. A Python Interface can run RPi encryption and decryption. Once the environment has been set up, a client connection can be established, and Python socket programming can be leveraged. TEA can secure parallel computing, which may increase system performance [34,51].

Implementing these methods gives a competitive advantage in protecting IoT devices. Pairing-based cryptography continues to make an objective impact, given its innovative and revolutionary security solutions. Despite these technological advances in modern security domains, IoT devices are susceptible to data breaches and DDoS attacks. Role Based Access Control (RBAC) is a standard way to allocate access based on user or program roles. Secure Encryption Based on Role (SEBR) is a way to grant access to encryption or decryption based on roles. This process allows different security layers to protect IoT devices and other systems based on the levels of encryption required. SEBR provides administrative and security control solutions to manage and protect data. The back-end and front-end applications are necessary to protect data provisioning and delivery. For proper protection, generated keys must be kept on the virtual or physical RPi-s [49].

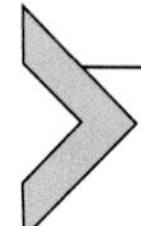

13. Hash authentication and integrated raspberry Pi capability

Physical RPi-s are affordable and can be inexpensively integrated into other systems. Virtual Raspberry Pi-s may be configured and deployed quickly. These virtual RPi-s can be managed on VMMs running on VT-x, AMD-V processors, or ARMs. If RPi-s are deployed in the enterprise, they may be helpful to validate or protect their identity, data, video files, and other contents.

Hardware characteristics must be shared with other endpoints in the assigned network as a node. In a perimeter-based configuration, hashes are likely unique signatures to establish, authenticate, and translate assigned signatures. Raspberry Pi carries out each process equipped with programming languages such as Python. Python is a general-purpose language designed for learning software development. Python is focused on ease of use. It is commonly applied to IoT networks. Delivering messages across other cloud platform applications can be done via a Firebase Cloud Messaging protocol. Android-based applications are helpful. Many Android devices are virtualizable on QEMU already. These can be installed on a remote Android device. Raspberry Pi pushes and receives digital notifications in case of a data breach or system compromise [52].

14. Motion passive infrared sensors for raspberry Pi

Surveillance systems may use infrared motion sensors, such as the HC-SR501 PIR. These systems may detect physical incursions or attacks and then activate CCTV systems. A virtual network of RPi-s can process and control the data gathering when supported by physical PIR and other cameras. The level of abstraction given by the virtualization of these RPi-s adds value by allowing migration, VM image storage and validation, ease of provisioning, fingerprinting provisions, and replication. These RPi-s can store recorded data and images online for future viewing or verification. RPi-s can keep validating fingerprints on a blockchain. Additionally,

Raspberry Pi-s have capabilities such as camera power reduction, cost efficiency, and durability [53].

It is possible to build more integrated motion detection and surveillance systems. Motion Passive Infrared Sensors (PI-Ss) are in use and can be installed in buildings or even on mobile systems. This chapter highlights motion detectors, alarms, and surveillance security systems. Sensors can be deployed and activated to provide security and protection to facilities. They offer an extra layer of security to property motion detectors, alarms, motion detectors, and surveillance systems. Virtual RPi-s add a level of abstraction and mobility for monitoring PI-Ss. PI-Ss are vital in the automobile, aerospace, technology, healthcare, banking, and related sectors [53].

15. Conclusion

This chapter focuses on cybersecurity applications on virtual Raspberry Pi-s. Virtual Raspberry Pi-s allow cybersecurity applications to be integrated into new enterprises more rapidly than physical RPi-s. Virtual RPi-s can be vertically or horizontally scaled quickly. The chapter discusses virtual RPi-s with a brief overview of surveillance monitoring and detection systems. It underscores how virtual RPi-s add to security. Deploying networks of virtual RPi-s in diverse locations allows them to gather security events, places and movements, and the general status of systems. Virtual Raspberry Pi-s provide advanced security options for isolation, migration, cloning, replication, and agility. These capabilities can augment security. Leveraging RPi-based blockchains can irrevocably store security events in RPi-based blockchain ledgers. This process can lead to following people and events while creating verifiable records of images, voices, and locations. Besides, it can lead to thwarting some deepfakes. Stopping deepfakes occurs when AI-based human image/synthetic media or deep learning methods cannot corroborate the date-time of the events they are claiming. We also discussed the Pi-hole messaging and tracking functions and the intrusion detection system for virtual Raspberry Pi-s. Finally, the chapter examines cybersecurity applications and virtual Raspberry Pi solutions. As an evolving technology, virtual Raspberry Pi can also be a trusted platform for building algorithmic models to generate text, images, and other data types. In addition, educators, developers, and researchers can use virtual Raspberry Pi to explore, innovate, and discover new technologies. Blockchain technology is a vital tool that users and organizations need to boost security, prevent theft of private keys, and minimize risk to digital cryptocurrencies. Virtual Raspberry Pi continues to play a critical role in blockchain technology. In distributed computing, devices like virtual Raspberry Pi have significantly impacted many services, including security.

References

[1] Y. Bai, H. Fan, X. Yang, J. Zhang, A blockchain-based reputation management model for enhancing cybersecurity in cloud computing, IEEE Trans. Serv. Comput. 13 (1) (2020) 90–101.

[2] A.S. Tanenbaum, A.S. Woodhull, Operating Systems Design and Implementation, Prentice Hall Press, 2015.

[3] F. Bellard, QEMU is a fast and portable dynamic translator, in: Proceedings of the Annual Conference on USENIX Annual Technical Conference, 2005, pp. 41–46.

[4] VMware, VMware Workstation Player, 2021, Retrieved from https://www.vmware.com/products/workstation-player.html.

[5] P.G. Bradford, D.A. Ray, Using digital chains of custody on constrained devices to verify evidence, in: Intelligence and Security Informatics (ISI 2007), New Brunswick, NJ, 2007.

[6] S. Kim, J. Lee, A virtual raspberry Pi environment for cybersecurity education and research, IEEE Trans. Learn. Technol. 13 (4) (2020) 589–597.

[7] C. Xu, X. Du, Integrating blockchain for data protection and cybersecurity: a case for Paychex, IEEE Technol Soc Magazine 37 (2) (2018) 57–64.

[8] "Raspberry Pi—Wikipedia," 2022. [Online]. Available: https://en.wikipedia.org/wiki/Raspberry_Pi. [Accessed 15 January 2022].

[9] C. Takemura, L.S. Crawford, The Book of Xen, No Starch Press, 2010.

[10] L. Zhang, D. Liu, Virtual raspberry Pi environments for natural language processing research, in: Proceedings of the IEEE 6th International Conference on Big Data Analytics, Los Angeles, CA, USA, 2020, pp. 290–295.

[11] A. Ramakrishnan, R. Chakraborty, Using virtual raspberry Pi environments for generative neural network research, in: Proceedings of the IEEE 18th International Conference on Machine Learning and Data Mining, Lisbon, Portugal, 2022, pp. 125–132.

[12] W.D. Ashley, Foundations of Libvirt Development, APress, 2019.

[13] "Raspberry Pi OS 2021-10-30 Release Notes," 2022. [Online]. Available: https://downloads.raspberrypi.org/raspios_armhf/release_notes.txt. [Retrieved: January, 2022].

[14] A. Smith, B. Johnson, The future of virtual raspberry Pi in machine learning, artificial intelligence, cybersecurity, and blockchain, in: IEEE Transactions on Emerging Topics in Computing, vol. 8, 2020, pp. 74–81. no. 1.

[15] R. Gupta, et al., Using virtual raspberry Pi environments for developing and testing AI and blockchain applications, in: Proceedings of the IEEE International Conference on Artificial Intelligence and Blockchain Technology, New York, NY, USA, 2019, pp. 126–133.

[16] G. Howser, Computer Networks and the Internet, in: Computer Networks and the Internet—A Hands-On Approach, Springer, 2020.

[17] S.K. Singh, S. Chawla, Security vulnerabilities and risks of raspberry Pi-based SCADA system, in: 2019 4th International Conference on Computing Methodologies and Communication (ICCMC), IEEE, 2019, https://doi.org/10.1109/ICCMC.2019.8725718.

[18] K. Bhargava, P.G. Bradford, N. Verba, Virtualizing IoT development, in: To Appear in the Proceedings of the American Society for Engineering Education (ASEE)-Northeast Conference, 2021.

[19] QEMU Team, QEMU (QEMU version 6.2.0 contains 2300+ commits from 189 authors), 2021. 18 12 2021. [Online]. Available: https://www.qemu.org/, [Retrieved: December, 2021].
[20] F. Bellard, QEMU, a Fast and Portable Dynamic Translator, in: ATEC '05: Proceedings of the Annual Conference on USENIX Annual Technical Conference, 2005 (Retrieved: December, 2021).
[21] Raspberry Pi Foundation, Raspberry Pi Operating System Images, Raspberry Pi OS Lite, 2021 (Online]. Available: https://www.raspberrypi.com/software/operating-systems/. [Retrieved: December, 2021).
[22] Raspberry Pi Foundation, Raspberry Pi, 2021. 19 12 2021. [Online]. https://www.raspberrypi.com/. [Retrieved: December, 2021].
[23] D.M. Gallagher, Analysis of Effects of Sensor Multithreading to Local System Event Timelines, Air Force Institute of Technology, 2014. Masters Thesis: AFIT-ENG-14-M-31.
[24] B.M. Jakobsson, Methods and Apparatus for Efficient Computation of One-way Chains in Cryptographic Applications, 2017. U.S. Patent US-9747458-B2, 29 08 2017.
[25] S. Nakamoto, Bitcoin: A Peer-to-Peer Electronic Cash System, 2008 (Online]. Available: https//bitcoin.org/bitcoin.pdf. [Accessed: 21-Apr-2023).
[26] A. Madan, V. Balaji, Blockchain-based Security for IoT Applications: A Review, in: Proceedings of the 2019 IEEE International Conference on Communication, Networks and Satellite (ComNetSat), Chennai, India, 3–5 October 2019, IEEE, 2019, pp. 1–5.
[27] S. Kumar, M. Sharma, A virtual raspberry Pi environment for cybersecurity education and training, IEEE Trans. Learn. Technol. 13 (4) (2020) 589–598.
[28] P.G. Bradford, D.A. Ray, Using digital chains of custody on constrained devices to verify evidence, IEEE Intell. Sec. Inform. (2007).
[29] M. Johnson, T. Lee, J. Smith, Virtual raspberry Pi with blockchains for cybersecurity applications, IEEE Trans Cybersec 8 (3) (2021) 245–255, https://doi.org/10.1109/TCYB.2021.3100567.
[30] S. Malhotra, 4 Next-gen IoT Applications With Raspberry PI, 2020 (Online]. https://artificialintelligence.oodles.io/blogs/iot-applications-with-raspberry-pi/. [Retrieved: August, 2020).
[31] E. Fernando, Meyliana and Surjandy, "Blockchain Technology Implementation In Raspberry Pi For Private Network,", Int. Conf. Sustain. Inform. Eng. Technol. (SIET) 2019 (2019) 154–158, https://doi.org/10.1109/SIET48054.2019.8986053.
[32] H. Razali, M.H. Alkawaz, A.S. Suhemi, Smart IoT Surveillance Multi-Camera Monitoring System, 2019, https://doi.org/10.1109/ICSPC47137.2019.9067984. https://www.researchgate.net/publication/340698618_Smart_IOT_Surveillance_Multi-Camera_Monitoring_System. (Retrieved: March 2022).
[33] NIST, Secure Hash Standard (SHS), 2015, Accessed: 28-Dec-2022. https://nvlpubs.nist.gov/nistpubs/fips/nist.fips.190-4.pdf.
[34] E. Conrad, J. Feldman, Next Generation SSH2 Implementation—Securing Data in Motion, 2012. https://www.sciencedirect.com/topics/computer-science/secure-hash-algorithm. Retrieved: March 2022.
[35] S.S. Sheshai, Raspberry Pi Based Security Systems PRJ Index 156, 2011. https://eie.uonbi.ac.ke/sites/default/files/cae/engineering/eie/RASPBERRY%20PI%20BASED%20SECURITY%20SYSTEM.pdf. (Retrieved: March, 2022).
[36] T. Smith, Virtual Raspberry PiL Software-Based Emulation, Instances, Operating System, Hardware, and Software, Physical and Virtual Machines, IEEE, 2021.
[37] M. Dai, X. Li, The advantages of virtual raspberry Pi in computer science education, IEEE Access 8 (2020) 125327–125334, https://doi.org/10.1109/ACCESS.2020.3008058.

[38] S. Al-Khalifa, M. Alshayeb, The potential of virtual raspberry Pi for cybersecurity education: advantages and challenges, in: Proceedings of the 2020 IEEE International Conference on Teaching, Assessment, and Learning for Engineering (TALE), Yokohama, Japan, 2020, pp. 138–143, https://doi.org/10.1109/TALE48000.2020.9368306.

[39] X. Gao, Z. Cai, Design and implementation of security education platform based on virtual raspberry Pi, in: 2018 International Conference on Computer, Information and Telecommunication Systems (CITS), IEEE, 2018, https://doi.org/10.1109/CITS.2018.8512142.

[40] C. Mulliner, G. Muehlberger, A raspberry Pi cluster for distributed pentesting and digital forensics, in: 2017 IEEE 41st Annual Computer Software and Applications Conference (COMPSAC), IEEE, 2017, https://doi.org/10.1109/COMPSAC.2017.45.

[41] Raspberry Pi Desktop, Raspberry Pi foundation, 2021. www.raspberrypi.org/software/raspberry-pi-desktop/.

[42] It's FOSS, Best Raspberry Pi Emulators to Choose When You Want to Try Out Raspbian, 2021. itsfoss.com/raspberry-pi-emulators.

[43] Cloudwards, Raspberry Pi on the Cloud: How to Set up Your Own Server, 2018. www.cloudwards.net/raspberry-pi-cloud/.

[44] Howchoo, 5 Raspberry Pi Virtual Machine Software Options, 2021. howchoo.com/g/5-raspberry-pi-virtual-machine-software-options.

[45] Raspberry Pi Foundation, Remote Access Your Raspberry Pi, 2021. www.raspberrypi.org/documentation/remote-access/.

[46] R.K. Dasgupta, Pi Black Hole for Internet Advertisements, 2018. https://www.researchgate.net/publication/326319875_Pi_Black_Hole_for_Internet_Advertisements. (Retrieved: February, 2022).

[47] K. Ranjeethapriya, N. Susila, G.R. Elwin, S. Balakrishnan, Raspberry Pi Based Intrusion Detection System, 2018. https://www.researchgate.net/publication/325952854_Raspberry_Pi_based_intrusion_detection_system?msclkid=81d4f7d2a7e711ec8bc3d053c232cedc. (Retrieved: March, 2022).

[48] B.S. Sunitha, A. Basu, SERBAC Framework on Encryption of Cloud Data Using Raspberry Pi, 2016. https://www.ijert.org/research/serbac-framework-on-encryption-of-cloud-data-using-raspberry-pi-IJERTCONV4IS29046.pdf. (Retrieved: March, 2022).

[49] N. Sharma, P.H. Sultana, R. Singh, S. Patil, Secure Hash Authentication in IoT Based Applications, 2019. https://reader.elsevier.com/reader/sd/pii/S1877050920300508?token=D80F6F5BE06687BDC895DE1A9155EF2EA304E060A2E63A2079706EB32E486A4C696891F76D2F884DF139BF2256549D5D&originRegion=us-east-1&originCreation=20220321173055. (Retrieved: March 2020).

[50] O.C. Nosiri, C.C. Akwiwu-Uzoma, U.A. Nmaju, C.H. Elumeziem, Motion Detector Security System for Indoor Geolocation, 2018. https://www.researchgate.net/publication/329876642_Motion_Detector_Security_System_for_Indoor_Geolocation. Retrieved: February 2022.

[51] M. El-hajj, M. Chamoun, A. Fadlallah, A. Serhrouchni, Analysis of Cryptographic Algorithms on IoT Hardware Platforms, 2019. https://www.researchgate.net/publication/330254752_Analysis_of_Cryptographic_Algorithms_on_IoT_Hardware_platforms/link/5c38884b299bf12be3bedbd1/download. Retrieved: March 2022.

[52] P. Blythe, J. Fridrich, Secure Digital Camera, in the proceedings of The Digital Forensic Research Conference DFRWS, 2004. USA.

[53] T. Powell, Encryption and Decryption with a RasPI Device, 2021. https://digitalcommons.odu.edu/cgi/viewcontent.cgi?article=1520&context=undergradsymposium. Retrieved: March 2021.

About the authors

Dr. Phillip Bradford is a computer scientist with extensive experience in academia and industry. He has a strong conviction about applying research to industrial challenges. He is currently in the Computer Science and Engineering faculty at the University of Connecticut. He worked for General Electric Asset Management, BlackRock, Reuters Analytics, and he co-founded a firm. He occasionally consults and often works with early-stage startups. He was on the faculty at the University of Alabama School of Engineering and at Rutgers Business School. He was a post-doctoral fellow at the Max-Planck-Institut für Informatik. He has several best-in-class results. He earned his PhD in computer science at Indiana University, an MS in computer science from the University of Kansas, and a BA in mathematics from Rutgers University.

Dr. Marcus Tanque is a principal researcher, professional, and scholar. He has worked with government-industry stakeholders. He is an independent researcher, author, editorial review board member, and referee for several books, journal articles, and published chapters. He holds a PhD in Information Technology with a dual Specialization in Information Assurance and Security and an MS in Information Systems Engineering. In his spare time, he enjoys helping and mentoring others. He is also an enthusiastic reader and researcher who collaborates with like-minded professionals.